Mathematics IIT JEE Main/Advanced - DPP

Bulls Eye - Sure Selection

Ramesh Chandra

B.Tech IIT Kanpur (Mech. Engineering)

Copyright © 2020 Ramesh Chandra

All rights reserved.

ISBN-13: 979–8-6548-7895-3

DEDICATION

To all hard-working students, who are putting their 100% to crack the competitive exam.

The Mathematics IIT JEE Main/Advanced - DPP is specially prepared for the students who are preparing for engineering entrance exam like jee main/Advanced, BITSAT etc. These daily practice problems are designed by 10 year teaching experienced senior professor Mr. Ramesh Chandra B.Tech IIT Kanpur (Mechanical Engineering). The book contains the higher order thinking problems, covering each and every concepts of the entire topic from level zero to advanced. Having this book add value to the aspirant's preparation.

Table of Contents

Squares & Square Roots — 9

___single option correct	9
___multiple options correct	10
___subjective Problems	10
___answer key – squares & square roots	12
_____single option correct, multi options correct, subjective	12

Fractions — 13

___single option correct	13
___multiple options correct	16
___integer type	17
___subjective Problems	18
___answer key - fractions	20
_____single option correct, multi options correct, integer type, subjective	20

Ratio & Proportion — 21

___single option correct	21
___multiple options correct	22
___subjective Problems	24
___answer key – ratio & proportion	27
_____single option correct, multi options correct, subjective	28

Algebraic Expressions and Identities — 29

___single option correct	29
___integer type	33
___subjective Problems	33
___answer key - algebraic expressions and identities	37
_____single option correct, integer type, subjective,	

Number System — 39

___single option correct	39
___multiple options correct	44
___integer type	46
___subjective problems	46

____answer key – number system	49
_____single option correct, multi options correct, integer type, subjective	49

Number System – NTSE   **50**

____single option correct	50
____integer type	53
____subjective problems	53
____answer key – number system ntse	54
_____single option correct, integer type, subjective	54

Introduction to Trigonometry – X   **55**

____section – a (brain nerves development)	55
____section – b (rank booster problems)	58
____section – c (in-depth analysis, brain storming problems)	59
____section – d (integer answer type)	61
____answer key – introduction to trigonometry	62

Modulus   **63**

____properties of $\|x\|$	63
____questions	64
_____Answer Key	67

Logarithm   **68**

____single option correct	68
____multiple options correct	70
____subjective problems	70
____logarithmic equations	72
____answer key - logarithm	73
_____single - multiple option correct, subjective problems, logarithmic equations	73

Sequences and Series   **74**

____single option correct	74
____multiple options correct	79
____integer type	82
____subjective problems	83
____matrix match	86
____answer key – sequence & series	88

____single - multi options correct, integer type, subjective, matrix match	89

Quadratic Equation & Inequations — 90

____single option correct	90
____multiple options correct	97
____integer type	98
____subjective problems	99
____matrix match	99
____comprehension for q 1 - 3	100
____answer key – quadratic equations & inequations	101
____single - multi options correct, integer type, subjective, matrix match, comprehension	102

Trigonometric Ratio and Identities — 103

____single option correct	103
____multiple options correct	107
____integer type	109
____matrix match	110
____comprehension	111
____answer key & solution	113
____single -multi options correct, integer type, matrix match, comprehension	113

Straight Lines — 114

____subjective problems (level - i)	114
____locus problems	120
____parametric form	121
____triangle points	122
____angle bisectors + family of lines	123
____subjective problems (level - ii)	124

Limits — 126

____single option correct	126
____paragraph q. no. 7 – 8	128
____answer key - limits	132

Differentiability — 133

____class work - understanding	133
____functional relationship	136

____answer key - differentiability	137
____functional relationship	137

Differentiation - Basic   138

____class work - understanding	138
____some more problems	140
_____answers to some more problems	142

Methods of Differentiation   143

____class work - understanding	143
_____answer key - differentiation	151

Matrix & Determinant   152

____single option correct	152
____multi option(s) correct	155
____integer option type (0 - 9)	157
____subjective problems	159
____answer key -matrix & determinant	162
_____single option correct, multi option correct, integer type, subjective problems	162

Relations & Functions   163

____section (a) n.c.e.r.t.	163
____section (b) jee-mains	165
____paragraph (question no. 11, 12, 13)	166
____section (c) jee-advance	167

Definite Integration   170

____single option correct	170
____multiple options correct	174
____integer type	175
____matrix match	176
____answer key – definite integration	177
_____single option correct, multi options correct, integer type	177

Vector & 3D   178

____single option correct	178
____multiple options correct	182
____integer type	183

____subjective problems	183
____answer key – vector & 3d	184
_____single option correct, multi options correct, integer type, subjective	184

Mathsarc Education

A learning place to fulfill your dream of success!

MATHEMATICS **IIT FOUNDATION**

SQUARES & SQUARE ROOTS

Master piece of an IITian author for competitive exams

SINGLE OPTION CORRECT

1. $2 \times 10^3 + 3 \times 10^2 + 5 \times 10$ equals

   (A) 235 (B) 2350 (C) 532 (D) None of these

2. Which of the following numbers is an odd integer, contains the digit 5, is divisible by 3, and lies between 12^2 and 13^2?

   (A) 105 (B) 147 (C) 156 (D) 165

3. If $x * y = x + y^2$, then $2 * 3$ equals

   (A) 8 (B) 25 (C) 11 (D) 13

4. When the numbers $\sqrt{5}$, 2.1, $\dfrac{7}{3}$, $2.05555\ldots\infty$, $2\dfrac{1}{5}$ are arranged in order from smallest to largest, the middle number is

   (A) $\sqrt{5}$ (B) 2.1 (C) 7/3 (D) $2\dfrac{1}{5}$

5. The number of natural numbers between 2313^2 and 2314^2

   (A) 4625 (B) 4626 (C) 4627 (D) 4635

6. Q is the point of intersection of the diagonals of one face of a cube whose edges have length 2 units. The length of QR is

   (A) 2 (B) $\sqrt{8}$ (C) $\sqrt{6}$ (D) $\sqrt{5}$

7. The number of integers between $-\sqrt{8}$ and $\sqrt{32}$ is

   (A) 9   (B) 7   (C) 6   (D) 8

8. In the diagram, square ABCD is made up of 36 squares, each with side length 1. The area of the square KLMN, in square units, is

   (A) 25   (B) 20   (C) 16   (D) 18

   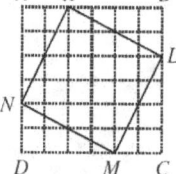

9. Which one is not a Pythagorean triplet?

   (A) (5, 12, 13)   (B) (20, 21, 29)   (C) (9, 40, 41)   (D) (11, 59, 60)

10. The value of $\sqrt{176 + \sqrt{2401}}$ is

    (A) 14   (B) 15   (C) 16   (D) 17

MULTIPLE OPTIONS CORRECT

1. Select the correct options

   (A) $3^2 + 4^2 + 12^2 = 13^2$ (B) $7^2 + 8^2 + 56^2 = 57^2$

   (C) $n^2 + (n+1)^2 + \{n(n+1)\}^2 = \{n(n+1)+1\}^2$ (D) $11^2 + 13^2 + 143^2 = 144^2$

2. Select the correct options

   (A) $1 + 3 + 5 + 7 + 9 + 11 = 6^2$ (B) $1 + 3 + 5 + 7 + \ldots + 51 = 25^2$

   (C) $1 + 3 + 5 + 7 + \ldots + 51 = 26^2$ (D) $1 + 3 + 5 + 7 + \ldots + 153 = 77^2$

3. Let $n \in N$ then select the Pythagorean triplet

   (A) $(2n, n^2 - 1, n^2 + 1)$ (B) $(2mn, m^2 - n^2, m^2 + n^2)$

   (C) $(4mn, m^2 - n^2, m^2 + n^2)$ (D) $\{2n+1, 2n(n+1), 2n(n+1)+1\}$

SUBJECTIVE PROBLEMS

1. Identify the prefect squares among the following numbers;

   1, 2, 3, 8, 36, 49, 65, 67, 71, 81, 169, 625, 125, 900, 100, 1000, 100000.

MATHSARC EDUCATION
IIT JEE MAINS / ADVANCED

2. Express 144 as a sum of 12 odd numbers.

3. Find the square root of the following numbers by factorisation:

   (i) 196   (ii) 256   (iii) 10404   (iv) 1156   (v) 13225.

4. Find the smallest positive integer with which one has to multiply each of the following numbers to get a perfect square:

   (i) 847   (ii) 450   (iii) 1445   (iv) 1352.

5. Find the smallest natural number by which 18900 must be divided so that the quotient is a perfect square. Find the square root of the quotient.

6. Find the smallest square number which is divisible by each of the numbers 12, 15 and 35.

7. Find the least number that must be subtracted from 57321 so as to get a perfect square. Also find the square root of the perfect square.

8. Find the least number that must be added to 5363 so as to get a perfect square. Also find the square root of the perfect square.

9. Write the square roots of following upto 3 place of decimals

   (i) 1203.569   (ii) 302.0131   (iii) 980.203   (iv) 5421.23

ANSWER KEY – Squares & Square Roots

SINGLE OPTION CORRECT

1. B
2. D
3. C
4. D
5. B
6. C
7. D
8. B
9. D
10. B

MULTI OPTIONS CORRECT

1. A, B, C
2. A, C, D
3. A, B, D

SUBJECTIVE

1. 36, 49, 81, 169, 625, 900 and 100
2. 1 + 3 + 5 +…..+ 23
3. (i) 14, (ii) 16, (iii) 102, (iv) 34, (v) 115
4. (i) 7   (ii) 2   (iii) 5   (iv) 2
5. 21, sqrt = 30
6. 44100
7. 200, 239
8. 113, 74
9. (i) 34.692 (ii) 17.378 (iii) 31.308 (iv) 73.629

Mathsarc Education

A learning place to fulfill your dream of success!

MATHEMATICS IIT FOUNDATION

FRACTIONS

An Idea can change your life in fraction of second!

SINGLE OPTION CORRECT

1. A square floor is tiled, as partially shown, with a large number of regular hexagonal tiles. The tiles are colored blue or white. Each blue tile is surrounded by 6 white tiles and each white tile is surrounded by 3 white and 3 blue tiles. Ignoring part tiles, the ratio of the number of blue tiles to the number of white tiles is closest to

    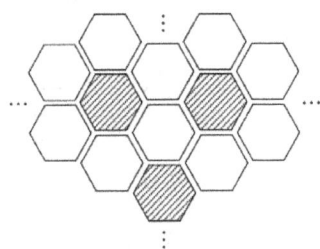

    (A) 1 : 6   (B) 2 : 3

    (C) 1 : 2   (D) 1 : 4

2. In the given diagram, all 12 of the small rectangles are the same size. Your task is to completely shade some of the rectangles until 2/3 of 3/4 of the diagram is shaded. The number of rectangles you need to shade is

    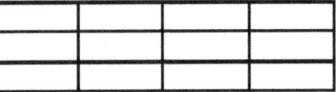

    (A) 9   (B) 3   (C) 4   (D) 6

3. The largest number in the set {0.01, 0.2, 0.03, 0.02, 0.1} is

    (A) 0.01   (B) 0.2   (C) 0.03   (D) 0.1

4. The value of 0.001 + 1.01 + 0.11 is

    (A) 1.111   (B) 1.101   (C) 1.013   (D) 1.121

5. If P = 1 and Q = 2, which of the following expressions is **not** equal to an integer?

    (A) P + Q   (B) P × Q   (C) P / Q   (D) P^Q

6. Four friends equally shared 3/4 of a pizza, which was left over after a party. What fraction of a whole pizza did each friend get?

   (A) 3/8   (B) 3/16   (C) 1/12   (D) 1/16

7. Which of the following fractions has the largest value?

   (A) 8/9   (B) 7/8   (C) 4/5   (D) 66/77

8. A small block is placed along a 10 cm ruler. Which of the following is closest to the length of the block?

   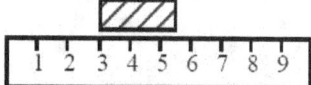

   (A) 0.24 cm   (B) 4.4 cm

   (C) 2.4 cm   (D) 3 cm

9. The sum $\dfrac{7}{10} + \dfrac{3}{100} + \dfrac{9}{1000}$ is equal to

   (A) 0.937   (B) 0.9037   (C) 0.7309   (D) 0.739

10. A square is divided, as shown. What fraction of the area of the square is shaded?

    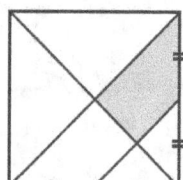

    (A) 1/4   (B) 1/8

    (C) 3/16   (D) 1/6

11. The smallest number in the set {3.2, 2.3, 3, 2.23, 3.22} is

    (A) 3.2   (B) 2.3   (C) 3   (D) 2.23

12. Which of the points positioned on the number line best represents the value of S ÷ T?

    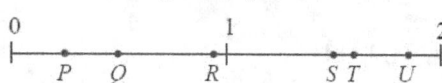

    (A) P   (B) Q   (C) R   (D) U

13. $\dfrac{1}{3} + \dfrac{1}{3} + \dfrac{1}{3} + \dfrac{1}{3} + \dfrac{1}{3} + \dfrac{1}{3} + \dfrac{1}{3}$ equals

    (A) $3\dfrac{1}{3}$ (B) $7 + \dfrac{1}{3}$ (C) $\dfrac{3}{7}$ (D) $7 \times \dfrac{1}{3}$

14. The largest fraction in the set $\left\{\dfrac{1}{2},\dfrac{1}{3},\dfrac{1}{4},\dfrac{1}{5},\dfrac{1}{10}\right\}$ is

    (A) $\dfrac{1}{2}$ (B) $\dfrac{1}{10}$ (C) $\dfrac{1}{3}$ (D) $\dfrac{1}{5}$

15. In the right - angled triangle PQR, PQ = QR. The segments QS, TU and VW are perpendicular to PR, and the segments ST and UV are perpendicular to QR, as shown. What fraction of PQR is shaded?

    (A) 3/16   (B) 3/8

    (C) 5/16   (D) 5/32

    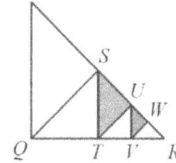

16. The diagram shown contains octagons and squares only. The ratio of the number of octagons to the number of squares is

    (A) 1 : 1   (B) 5 : 4

    (C) 5 : 3   (D) 25 : 12

    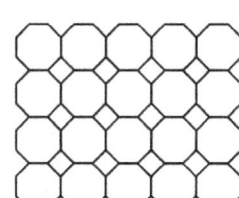

17. The value of: $\dfrac{20092008^2}{20092007^2 + 20092009^2 - 2}$ is

    (A) 1   (B) 2/3   (C) 1/2   (D) None of these

18. The rectangle PQRS is divided into six equal squares and shaded as shown. What fraction of PQRS is shaded?

    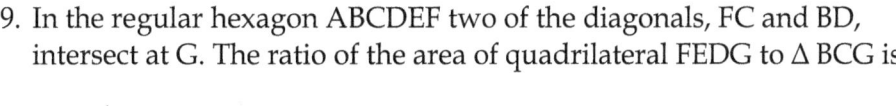

    (A) 1/2   (B) 7/12   (C) 5/12   (D) 5/11

19. In the regular hexagon ABCDEF two of the diagonals, FC and BD, intersect at G. The ratio of the area of quadrilateral FEDG to △ BCG is

    (A) $3\sqrt{3}$: 1   (B) 4 : 1

    (C) $2\sqrt{3}$: 1   (D) 5 : 1

    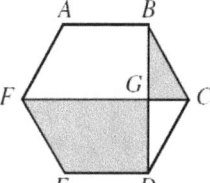

20. Simplify: $9\frac{2}{3} - 23\frac{1}{2} - 16\frac{3}{4} + 35\frac{5}{6}$

   (A) $5\frac{1}{4}$ (B) $4\frac{3}{4}$ (C) $-3\frac{1}{4}$ (D) None of these

21. The value of $1 + \cfrac{1}{2 + \cfrac{1}{3 + \cfrac{1}{4}}}$ is

   (A) $\frac{1}{10}$ (B) $1\frac{13}{30}$ (C) $1\frac{1}{10}$ (D) None of these

22. If $\frac{a}{b} = \frac{2}{3}$ and $\frac{b}{c} = \frac{4}{5}$, then value of $\frac{a+b}{b+c}$ is

   (A) $\frac{5}{9}$ (B) $\frac{20}{27}$ (C) $\frac{25}{27}$ (D) $\frac{6}{8}$

MULTIPLE OPTIONS CORRECT

1. Select the correct statements for integers a, b, c, d, e, f.

   (A) If a, b, c, d are integers such that b and d are positive and $\frac{a}{b}$ and $\frac{c}{d}$ fractions, then the fraction $\frac{a+c}{b+d}$ lies in between $\frac{a}{b}$ and $\frac{c}{d}$.

   (B) If $\frac{a}{b}, \frac{c}{d}, \frac{e}{f}, \ldots$ are equal fractions, with b, d, f, ... positive integer then the fraction $\frac{a+c+e+\ldots}{b+d+f+\ldots}$ is equal to all of them.

   (C) If a, b, c, d are integers such that b and d are positive and $\frac{a}{b} = \frac{c}{d}$, then $\frac{a}{b} = \frac{a + k \cdot c}{b + k \cdot d} = \frac{c}{d}$, where $k \in Q$.

   (D) If $\frac{a}{b} = \frac{c}{d}$, then $\frac{a+b}{a-b} = \frac{c+d}{c-d}$.

2. Select the correct option(s)

   (A) $\dfrac{2}{3}$ is a proper fraction

   (B) $\dfrac{7}{5}$ is an improper fraction

   (C) $3\dfrac{4}{5} = 3 + \dfrac{4}{5} = \dfrac{19}{5}$

   (D) $7\dfrac{3}{13}$ is a mixed fraction

3. Select the correct options

   (A) Like fractions have same denominator

   (B) Unlike fraction fractions have different Numerators

   (C) $\dfrac{5}{7} > \dfrac{3}{7} > \dfrac{2}{7}$ i.e. in like fraction, greater Numerator will have more value on number line

   (D) $\dfrac{6}{7} > \dfrac{5}{6}$, as 6×6 > 7×5.

INTEGER TYPE

1. Evaluate the expression: $\dfrac{1^2}{1^2 - 10 + 50} + \dfrac{2^2}{2^2 - 20 + 50} + \ldots\ldots + \dfrac{9^2}{9^2 - 90 + 50}$

2. Solve: $\dfrac{1}{9}\left\{\dfrac{1}{7}\left[\dfrac{1}{5}\left(\dfrac{x+2}{3} + 4\right) + 6\right] + 8\right\} = 1$

3. The value of: $1 - \dfrac{1}{1 - \dfrac{1}{1 - \dfrac{1}{2}}}$

4. Solve the equation $\dfrac{1}{10}\left\{\dfrac{1}{9}\left[\dfrac{1}{5}\left(\dfrac{x+2}{3} + 8\right) + 16\right] + 8\right\} = 1$

MATHSARC EDUCATION
A learning place to fulfill your dreams of success!

IIT JEE MAINS / ADVANCED

SUBJECTIVE PROBLEMS

1. Evaluate $(-5)^2 \times \left(-\dfrac{1}{5}\right)^3 - 2^3 \div \left(-\dfrac{1}{2}\right)^2 - (-1)^{1999}$.

2. Evaluate the following:

   (i) $(2 \times 3 \times 5 \times 7)\left(\dfrac{1}{2} + \dfrac{1}{3} + \dfrac{1}{5} + \dfrac{1}{7}\right)$

   (ii) $(-0.125)^7 \times 8^8$

   (iii) $(-11) + (-33) - (-55) - (-66) - (-77) - (-88)$

   (iv) $\left(-\dfrac{75}{13}\right)^2 + \left(\dfrac{37}{13}\right)^2$

   (v) $\left[\left(-\dfrac{6}{7}\right)^7 + \left(-\dfrac{4}{5}\right) \times \left(-\dfrac{4}{9}\right) \times \dfrac{16}{81}\right] \times \left(9\dfrac{246}{247} - 0.666\right)$

3. The value of $\dfrac{13579}{(-13579)^2 + (-13578)(13580)}$ is _____

4. Simplify: $\dfrac{83^3 + 17^3}{83 \times 66 + 17^2}$ _____

5. The value of $3 - \dfrac{1}{2} - \dfrac{1}{6} - \dfrac{1}{12} - \dfrac{1}{20} - \dfrac{1}{30} - \dfrac{1}{42} - \dfrac{1}{56} =$ _____

6. Evaluate: $\dfrac{1}{3} + \dfrac{1}{15} + \dfrac{1}{35} + \dfrac{1}{63} + \dfrac{1}{99} + \dfrac{1}{143}$.

7. If $-1 < a < 0$, then arrange the following in ascending order

   $a^3, -a^3, a^4, -a^4, \dfrac{1}{a}, -\dfrac{1}{a}$.

8. Let a and b be positive integers such that $\dfrac{2}{3} < \dfrac{a}{b} < \dfrac{5}{6}$. Find the value of a + b when b has the minimum value.

9. Plot the following on number line

   (i) 3/10 (ii) 4/5 (iii) $2\dfrac{3}{4}$ (iv) 1.3

10. Simplify the following:

(i) $27\dfrac{1}{2} + 15\dfrac{3}{4} - 12\dfrac{2}{5} + 18\dfrac{4}{5}$

(ii) $13\dfrac{3}{4} + 17\dfrac{2}{7} + 31\dfrac{1}{4} + 15\dfrac{5}{7} + 12\dfrac{2}{3}$

(iii) $9\dfrac{2}{3} - 23\dfrac{1}{2} - 16\dfrac{3}{4} + 35\dfrac{5}{6}$

11. Why $\dfrac{1}{2} = \dfrac{2}{4} = \dfrac{3}{6} = \ldots$ are equivalent fractions, Explain?

12. Simplify: $\dfrac{3x}{4} - \dfrac{x+2}{3} + \dfrac{x+2}{12}$

13. Solve the equation: $\dfrac{1}{5}\left\{\dfrac{1}{4}\left[\dfrac{1}{3}\left(\dfrac{x}{2} - 3\right) - 2\right] - 1\right\} - 2 = 1$

14. Solve the equation: $1 - \dfrac{x - \dfrac{1+3x}{5}}{3} = \dfrac{x}{2} - \dfrac{2x - \dfrac{10-6x}{7}}{2}$

Mathsarc Education
A learning place to fulfill your dreams of success!

THANKS!

Keep smiling!

Visit Us: https://www.mathsarc.com

ANSWER KEY - Fractions

SINGLE OPTION CORRECT

1. C
2. D
3. B
4. D
5. C
6. B
7. A
8. C
9. D
10. C
11. D
12. C
13. D
14. A
15. D
16. C
17. C
18. C
19. D
20. A
21. B
22. B

MULTI OPTIONS CORRECT

1. A, B, C, D
2. A, B, C, D
3. A, C, D

INTEGER TYPE

1. 9
2. 1
3. 2
4. 4

SUBJECTIVE

1. $-31\dfrac{1}{5}$
2. (i) 247 (ii) -8 (iii) 242 (iv) 45
3. 13579
4. 100
5. $2\dfrac{1}{8}$
6. 6/13
7. $\dfrac{1}{a} < a^3 < -a^4 < a^4 < -a^3 < -\dfrac{1}{a}$
8. 16
10. (i) $49\dfrac{13}{20}$ (ii) $90\dfrac{2}{3}$ (iii) $5\dfrac{1}{4}$
12. $\dfrac{x-1}{2}$
13. $x = 402$
14. $x = -\dfrac{74}{167}$

Mathsarc Education

A learning place to fulfill your dream of success!

MATHEMATICS **IIT FOUNDATION**

RATIO & PROPORTION

Your rewards in life are in direct proportion to your service!

SINGLE OPTION CORRECT

1. If $x : y : z = 2 : 3 : 4$, then ratio $\dfrac{x^3 + y^3 - 3xyz}{(x+y+z)^3}$ is

   (A) $1 : 9$ (B) $1 : 27$ (C) $2 : 27$ (D) $-\dfrac{37}{729}$

2. If $\dfrac{x+y}{2} = \dfrac{y+z}{3} = \dfrac{z+x}{4}$, then $x : y : z$ is

   (A) $1 : 3 : 5$ (B) $3 : 5 : 1$ (C) $3 : 1 : 5$ (D) $2 : 3 : 5$

3. If $(x - 2) : (3x - 29)$ be the duplicate ratio of $4 : 5$ then value of x is

   (A) 18 (B) 28 (C) 16 (D) None of these

4. If $(x + y) : (x - y) = 7 : 3$ then the ratio $(x^3 + y^3) : (x^3 - y^3)$ is

   (A) $133 : 117$ (B) $13 : 17$ (C) $7^3 : 3^3$ (D) None of these

5. The ratio of income of A and B is $3 : 4$. The ratio of their expenditure is $4 : 5$. Find the ratio of their savings if the savings of A is one fourth of his income

   (A) $3 : 5$ (B) $11 : 13$ (C) $12 : 17$ (D) $12 : 19$

6. the dimensions of a rectangular room $(l \times w)$ m² when increased by 4 meter, are in the ratio $4 : 3$ and when decreased by 4 meters are in the ratio of $2 : 1$ then

   (A) $l = 12, w = 8$ (B) $l = 8, w = 12$ (C) $l = 24, w = 16$ (D) $l = 16, w = 24$

7. In a mixture of three varieties of tea, the ratio of their weights is 4 : 5 : 8. If 5 kg tea of first variety, 10 kg of the second variety and some quantity of tea of the third variety are added to the mixture, the ratio of the weights of three varieties of tea becomes 5 : 7 : 9. The quantity of third variety of tea in the final mixture is ___

   (A) 40 kg (B) 45 kg (C) 50 kg (D) 35 kg

8. Initially two cup of same volume are filled with milk upto $\frac{3}{5}$ th and $\frac{4}{5}$ th of their volume. Water is then filled. Then two mixtures are mixed and poured into a jug. The ratio of water to milk in the mixture is _

   (A) 3 : 5 (B) 2 : 3 (C) 3 : 7 (D) 3 : 4

9. Let $\frac{a}{b} : -\frac{b}{a} = x : y$. If $(x - y) = \left\{\frac{a}{b} + \frac{b}{a}\right\}$, then x is equal to ___

   (A) $\frac{a-b}{a}$ (B) $\frac{a+b}{a}$ (C) $\frac{a+b}{b}$ (D) None of these

10. If p : q : r = 1 : 2 : 3, then $\sqrt{12p^2 + q^2 + r^2}$ is equals to

    (A) 5 (B) 2q (C) 5p (D) 4r

11. If A varies directly proportional to C and B also varies directly proportional to C, which one of the following is not correct?

    (A) $(A + B) \propto C$ (B) $(A - B) \propto \frac{1}{C}$ (C) $\sqrt{(AB)} \propto C$ (D) $\frac{A}{B}$ = Constant

12. Third proportion to 25 and 30 is

    (A) 36 (B) 32 (C) 34 (D) 38

13. If m : n = 3 : 2 then value of (4m + 5n) : (4m – 5n) is ___

    (A) 5 (B) 7 (C) 13 (D) 11

MULTIPLE OPTIONS CORRECT

1. If $y(3x - y) : x(4x + y) = 5 : 12$, then the ratio $(x^2 + y^2) : (x + y)^2$ is

   (A) 4/5 (B) 25/49 (C) 41/81 (D) 3/4

2. If $\dfrac{a}{b+c} = \dfrac{b}{c+a} = \dfrac{c}{a+b}$ then this ratio could be

   (A) 1/3 (B) 1/2 (C) 1 (D) -1

3. If $\dfrac{a}{b} = \dfrac{c}{d}$ then select the correct options

   (A) a d = b c
   (B) $\dfrac{a}{c} = \dfrac{b}{d}$
   (C) $\dfrac{a}{b} = \dfrac{c}{d} = \dfrac{a+c}{b+d}$
   (D) $\dfrac{a}{b} = \dfrac{c}{d} \to \dfrac{a+b}{a-b} = \dfrac{c+d}{c-d}$

4. Given that $\dfrac{a}{b} = \dfrac{2}{3}$, which of the following is/are not a true statement?

   (A) a = 2, b = 3
   (B) a = 2k, b = 3k (k ≠ 0)
   (C) 3a = 2b
   (D) $a = \dfrac{2}{3} b$

5. Which of the following is/are true statements,

   (A) if $\dfrac{a}{b} = \dfrac{c}{d}$, then $\dfrac{a}{b} = \dfrac{c+m}{d+m}$
   (B) if $\dfrac{a}{b} = \dfrac{c}{d}$, then $\dfrac{a^2}{b^2} = \dfrac{c^2}{d^2}$
   (C) if $\dfrac{a}{b} = \dfrac{c}{d}$, then $\dfrac{a+c}{b} = \dfrac{b+c}{d}$
   (D) if $\dfrac{a^2}{b^2} = \dfrac{c^2}{d^2}$, then $\dfrac{a}{b} = \dfrac{c}{d}$ or $\dfrac{a}{b} = -\dfrac{c}{d}$

6. If $6x^2 - 5xy + y^2 = 0$, where (x, y ≠ 0), then the value of $\dfrac{2x - 3y}{2x + 3y}$ is

   (A) 5/7 (B) -7/11 (C) -1/2 (D) 1/2

7. The numbers are in the ratio 7 : 3. If 10 is added to each number the ratio is halved. The numbers are

   (A) 10/3 (B) 10/7 (C) 10/21 (D) 20/7

8. If $y^2 - 3xy + 2x^2 = 0$ then x : y is equal to ____

   (A) 1/2 (B) 2/3 (C) 1/3 (D) 1

9. If x : 2 = y : 3 and 6x + 7y = 11 then

   (A) y = 2 (B) y = 1 (C) $x = \frac{2}{3}$ (D) x = 5/3

10. Select the correct statement(s)

    (A) Duplicate ratio of a : b is $a^2 : b^2$

    (B) sub-duplicate ratio of a : b is $\sqrt{a} : \sqrt{b}$

    (C) Triplicate ratio of a : b is $a^3 : b^3$

    (D) Sub - triplicate ratio of a : b is $\sqrt[3]{a} : \sqrt[3]{b}$

11. Select the correct statement(s)

    (A) If a : b :: c : d then d is fourth proportion

    (B) If a : b :: b : c then c is called third proportion to a and b

    (C) if c is the mean proportion of a : b then $c = \sqrt{ab}$ or a : c :: c : b

    (D) if a : b :: c : d then ac = bd

12. Select the correct statement(s)

    (A) If x is directly proportional to y ie. $x \propto y$ then x = ky, where k is proportionality constant

    (B) If x is inversely proportional to y ie. $x \propto \frac{1}{y}$ then xy = k, where k is proportionality constant

    (C) If $\frac{x}{y} = \frac{2}{3}$ then x = 2 and y = 3

    (D) If $\frac{a^2}{b^2} = \frac{4}{9}$ then $\frac{a}{b} = \frac{2}{3}$

SUBJECTIVE PROBLEMS

1. If A : B = 6 : 7 & B : C = 8 : 9 then find A : B : C

2. what will be the fraction whose ratio with 1/27 is equal to the ratio of 3/11 and 5/9?

3. If a : b = 2 : 3, b : c = 5 : 4, c : d = 3 : 7, then find a : b : c : d.

4. If 2x = 3y = 4z then find x : y : z

5. If a : b = 4 : 5 and b : c = 5 : 3 then find (i) a : c (ii) a : b : c

6. If 4A = 5B and 3 B = 5 C then find (i) A : C (ii) A : B : C

7. If A : B = 2 : 3, B : C = 3 : 2, C : D = 4 : 3 then find (i) A : D (ii) A : B : C : D

8. If $A : B = \dfrac{2}{3} : \dfrac{4}{5}$, $B : C = \dfrac{1}{4} : \dfrac{1}{5}$ and $C : D = \dfrac{2}{4} : \dfrac{1}{5}$ then (i) A : D = ? (ii) A : B : C : D = ?

9. If A : B = 3 : 2, B : C = 4 : 3, C : D = 1 : 2 and D : E = 1 : 2 then find (i) A : E (ii) A : B : C : D : E

10. If $a : b = \dfrac{2}{9} : \dfrac{1}{3}$, $b : c = \dfrac{2}{7} : \dfrac{5}{14}$ and $d : c = \dfrac{7}{10} : \dfrac{3}{5}$ then find a : b : c : d

11. If x : y = 5 : 2 then find the value of following:

    (i) $\dfrac{x^3 - y^3}{x^3 + y^3}$ (ii) $\dfrac{x^2 - xy + y^2}{x^2 + xy + y^2}$ (iii) $\dfrac{xy - x^2}{x^2 + y^2}$

12. If 2A = 3B and 4 B = 2C then find (i) $\dfrac{3A + 4B + 5C}{4A + 2B - 2C}$ (ii) $\dfrac{3A^2 + 4B^2 + 5C^2}{3A^2 + 4C^2}$

13. If a : b = 3 : 4 and b : c = 4 : 7 then value of (i) $\dfrac{a+b+c}{c}$ (ii) $\dfrac{a+c}{b} : \dfrac{a+b}{c}$

14. If (4x² – 3y²) : (2x² + 5y²) = 12 : 19 then find x : y

15. If $\dfrac{a}{b} = \dfrac{c}{d} = \dfrac{3}{2}$ then value of $\dfrac{13a - 111c}{13b - 111d}$ is

16. If $\dfrac{a}{b} = \dfrac{c}{d} = \dfrac{e}{f} = 3$ then find the value of

    (i) $\dfrac{ma + nc + pe}{mb + nd + pf}$ (ii) $\dfrac{ma^2 + nc^2 + pe^2}{mb^2 + nd^2 + pf^2}$ (iii) $\dfrac{m^2a + n^2c + p^2e}{m^2b + n^2d + p^2f}$

17. If x, 5, 10, y are in continued proportion, then find x and y

18. The ratio of the sum of money in three bags A, B and C is 4 : 3 : 2. If Rs 50 is added to each of the bags, the ratio becomes 14 : 13 : 12. Find the sum of money in each of the bags.

19. If (5x + 3) : (3x + 1) is the triplicate ratio of 4 : 3, then find x.

20. If $\dfrac{x}{a} = \dfrac{y}{b} = \dfrac{z}{c}$, then prove that $\dfrac{x^3}{a^3} + \dfrac{y^3}{b^3} + \dfrac{z^3}{c^3} = \dfrac{3xyz}{abc}$.

21. When the price of rice increases $12\dfrac{1}{2}$ % a man can get 250 gm less rice for Rs 18. Find the present cost of rice per kg.

22. If $\dfrac{a}{b} = \dfrac{c}{d} = \dfrac{e}{f}$, then show that $\dfrac{a^3b + 2c^2e - 3ae^2f}{b^4 + 2d^2f - 3bf^3} = \dfrac{ace}{bdf}$

23. If $P = \dfrac{x^2 - 36}{x^2 - 49}$ and $Q = \dfrac{x+6}{x+7}$, then find the value of P/Q.

24. If $\dfrac{x}{3} = \dfrac{y}{4} = \dfrac{z}{5}$, and $4x - 5y + 2z = 10$, find the value of $2x - 5y + z$.

25. Given $\dfrac{x}{y} = 2$, find the value of $\dfrac{2x^2 - 3xy + y^2}{x^2 + 2y^2}$

26. Given that $x : y = 3 : 5$ and $y : z = 2 : 3$, find the value of $\dfrac{x+y-z}{2x-y+z}$

27. Given that $2x - 3y + z = 0$, $3x - 2y - 6z = 0$ and $xyz \neq 0$, find the value of

   (i) $x : y : z$ (ii) $\dfrac{x^2 + y^2 + z^2}{2x^2 + y^2 - z^2}$

28. Two brothers have their ages in the ratio 3 : 4. In twelve years, their ages will be in the ratio 5 : 6. What are their ages now?

29. The angles of a triangle are in the ratio 2 : 5 : 11. Find them ____

30. The angles of a quadrilateral are in the ratio 1 : 2 : 3 : 4. What are the angles?

THANKS!

☺

Keep smiling!

Visit Us: https://www.mathsarc.com

ANSWER KEY – Ratio & Proportion

SINGLE OPTION CORRECT

1. D   2. C   3. A   4. A
5. D   6. A   7. B   8. C
9. D   10. C   11. B   12. A
13. D

MULTI OPTIONS CORRECT

1. B, C   2. B, D   3. A, B, C, D   4. A, B
5. B, D   6. B, C   7. A, B   8. A, D
9. B, C   10. A, B, C, D   11. A, B, C   12. A, B

SUBJECTIVE

1. 48 : 56 : 63
2. 1/55
3. 10 : 15 : 12 : 28
4. 6 : 4 : 3
5. (i) 4 : 3   (ii) 4 : 5 : 3
6. (i) 25 : 12   (ii) 25 : 20 : 12
7. (i) 4 : 3   (ii) 4 : 6 : 4 : 3
8. (i) 125 : 48   (ii) 125 : 150 : 120 : 48
9. 6 : 4 : 3 : 6 : 12
10. 16 : 24 : 30 : 35
11. (i) 117 : 133   (ii) 19 : 39   (iii) – 15 : 29
12. (i) 37 : 8   (ii) 123 : 91
13. (i) 2   (ii) 5 : 2
14. 3 : 2
15. 3 : 2
16. (i) 3   (ii) 9   (iii) 3
17. $x = 5/2$, $y = 20$
18. Rs 20, Rs 15, Rs 10
19. 17/57
21. 9
23. $\dfrac{x-6}{x-7}$
24. – 45
25. 1/2
26. 1/17
27. (i) 4 : 3 : 1   (ii) 13 : 20
28. 18 and 24
29. 20°, 50° and 110°
30. 36°, 72°, 108° and 144°

Mathsarc Education

A learning place to fulfill your dream of success!

MATHEMATICS   **IIT FOUNDATION**

ALGEBRAIC EXPRESSIONS AND IDENTITIES

Master piece of an IITian author for competitive exams

SINGLE OPTION CORRECT

1. If $x = -2$, the value of $(x)(x^2)\left(\dfrac{1}{x}\right)$ is

   (A) 4   (B) $-8\dfrac{1}{2}$ (C) -4 (D) 16

2. If $x = 3$, which of the following expressions is an even number?

   (A) $9x$ (B) x^3 (C) $2(x^2 + 9)$ (D) $2x^2 + 9$

3. If $2x - 1 = 5$ and $3y + 2 = 17$, then the value of $2x + 3y$ is

   (A) 8   (B) 19   (C) 21   (D) 23

4. If $y = 6 + \dfrac{1}{6}$, then $\dfrac{1}{y}$ is

   (A) 6/37   (B) 37/6   (C) 6/7   (D) 7/6

5. The least value of x which makes $\dfrac{24}{x-4}$ an integer is

   (A) -44 (B) -28 (C) -20 (D) 0

6. The 50th term in the sequence $5, 6x, 7x^2, 8x^3, 9x^4, \ldots$ is

   (A) $54\,x^{49}$ (B) $54\,x^{50}$ (C) $45\,x^{49}$ (D) $46\,x^{51}$

7. ABCD is a square with AB = x +16 and BC = 3x, as shown.

   The perimeter of ABCD is

   (A) 16
   (B) 32
   (C) 96
   (D) 48

   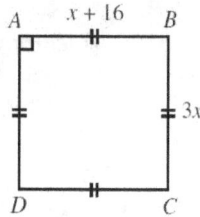

8. For how many integer values of x is $\sqrt{25-x^2}$ equal to an integer?

   (A) 7
   (B) 6
   (C) 5
   (D) 3

9. $3a + \{-4b - [4a - 7b - (-4a - b)] + 5a\}$ simplify to

   (A) 2a
   (B) 2b
   (C) 3a
   (D) 3b

10. Given $x^2 - x - 1 = 0$, simplify $\dfrac{x^3 + x + 1}{x^5}$ to a polynomial form.

    (A) 2x + 1
    (B) x + 1
    (C) x - 1
    (D) None of these

11. If $\dfrac{1}{x} - \dfrac{1}{y} = 4$, the value of $\dfrac{2x + 4xy - 2y}{x - y - 2xy}$ is

    (A) 2/3
    (B) 3/2
    (C) 3/4
    (D) 4/3

12. Suppose a, b are two real numbers such that $a^2 + b^2 + 8a - 14b + 65 = 0$, then value of

    $a^2 + ab + b^2$ is

    (A) 65
    (B) 37
    (C) 21
    (D) None of these

13. If the sum of two numbers is 22 and sum of their squares is 404 then the product of the number is

    (A) 40
    (B) 44
    (C) 80
    (D) 88

14. If $1^3 + 2^3 + 3^3 + \ldots + 10^3 = 3025$ then $4 + 32 + 108 + \ldots + 4000$ is equal to

    (A) 1200
    (B) 12100
    (C) 12200
    (D) 12400

15. The expression $(1+q)(1+q^2)(1+q^4)(1+q^8)(1+q^{16})(1+q^{32})(1+q^{64})$ where $q \neq 1$, equals

    (A) $\dfrac{1-q^{128}}{1-q}$
    (B) $\dfrac{1-q^{64}}{1-q}$
    (C) $\dfrac{1-q^{2^{1+2+3+\ldots+6}}}{1-q}$
    (D) none of these

16. Consider the statement: $x(\alpha - x) < y(\alpha - y)$ for all x, y with $0 < x < y < 1$. The statement is true

    (A) If and only if $\alpha \geq 2$
    (B) If and only if $\alpha > 2$
    (C) If and only if $\alpha < -1$
    (D) for no values of α

17. The number of ordered pairs of integers (x, y) satisfying the equation $x^2 + 6x + y^2 = 4$ is

    (A) 2
    (B) 4
    (C) 6
    (D) 8

18. The number of integer (positive, negative or zero) solutions of $xy - 6(x + y) = 0$ with $x \leq y$ is

    (A) 5
    (B) 10
    (C) 12
    (D) 9

19. If n is a positive integer such that $8n + 1$ is a perfect square, then

    (A) n must be odd
    (B) n cannot be a perfect square
    (C) n must be a prime number
    (D) 2n can not be a perfect square

20. Consider the following equation in x and y:

    $(x - 2y - 1)^2 + (4x + 3y - 4)^2 + (x - 2y - 1)(4x + 3y - 4) = 0$.

    How many solutions to (x, y) with x, y real, does the equation have?

    (A) none
    (B) exactly one
    (C) exactly two
    (D) more than two

21. Let a, b and c be distinct real numbers such that $a^2 - b = b^2 - c = c^2 - a$. Then $(a + b)(b + c)(c + a)$ equals

    (A) 0
    (B) 1
    (C) -1
    (D) none of these

22. x and y are two non-negative numbers such that $2x + y = 10$. The sum of the maximum and minimum values of $(x + y)$ is

    (A) 6
    (B) 9
    (C) 10
    (D) 15

23. If m = n² - n, where n is an integer, then m² - 2m is divisible by

    (A) 20　　　　　(B) 24　　　　　(C) 30　　　　　(D) 16

24. If $x = \dfrac{3+\sqrt{5}}{2}$ and y = x³, then y satisfies the quadratic equation

    (A) y² – 18y + 1 = 0　　(B) y² + 18y + 1 = 0　　(C) y² – 18y - 1 = 0　　(D) y² + 18y - 1 = 0

25. When a natural number x is divided by 5, the remainder is 2. When a natural number y is divided by 5, the remainder is 4. The remainder is z when x + y is divided by 5. The value of $\dfrac{2z-5}{3}$ is

    (A) - 1　　　　(B) 1　　　　(C) - 2　　　　(D) 2

26. If x + y + z = 1, xy + yz + zx = - 1, xyz = - 1 then value of x³ + y³ + z³ is:

    (A) - 1　　　　(B) 1　　　　(C) 2　　　　(D) - 2

27. The value of $\left(a^{1/8} + a^{-1/8}\right)\left(a^{1/8} - a^{-1/8}\right)\left(a^{1/4} + a^{-1/4}\right)\left(a^{1/2} + a^{-1/2}\right)$ is

    (A) $a + a^{-1}$　　(B) $a - a^{-1}$　　(C) $a^2 - a^{-2}$　　(D) $a^{1/2} - a^{-1/2}$

28. If ab < 0, then the relation in sizes of (a - b)² and (a + b)² is

    (A) $(a-b)^2 < (a+b)^2$　　　　　(B) $(a-b)^2 = (a+b)^2$

    (C) $(a-b)^2 > (a+b)^2$　　　　　(D) not determined

29. Let $\dfrac{a^3 + b^3 + c^3 - 3abc}{a+b+c} = 3$, then the value of (a - b)² + (b - c)² + (a - b)(b - c) is

    (A) 1　　　　(B) 2　　　　(C) 3　　　　(D) 4

30. The expression 4 {(3x - 2) - [3(3x - 2) + 3]} - (4 - 6x) simplifies to

    (A) - 18x　　　(B) 9x　　　(C) - 16 x　　　(D) None of these

INTEGER TYPE

1. Given $a^4 + a^3 + a^2 + a + 1 = 0$. Find the value of $a^{2000} + a^{2010} + 1$.

2. Given that the equation $2a(x + 6) = 4x + 1$ has no solution, where a is a parameter, find the value of a.

3. Given that the equation $ax + 4 = 3x - b$ has more than 1 solution for x.

   Find the value of $(4a + 3b)^{2007}$.

4. The number of integral solutions of the equation $7\left(y + \dfrac{1}{y}\right) - 2\left(y^2 + \dfrac{1}{y^2}\right) = 9$ is/are ___

5. The value of $\dfrac{a^3 + b^3 + c^3 - 3abc}{ab + bc + ca - a^2 - b^2 - c^2}$, when a = - 5, b = - 6, c = 10 is ___

6. If ab = 1, find the value of $\dfrac{a}{a+1} + \dfrac{b}{b+1}$.

7. If $x + y = 2z$, $x \neq y$, then find the value of $\dfrac{x}{x-z} + \dfrac{y}{y-z}$.

8. If $a \in Q$, find the least value of $(-a) + |a| + |-a| + (-a)$.

9. Given that $a - b = 2$, $b - c = -3$, $c - d = 5$, find the value of $(a - c)(a - d) \div (d - b)$.

SUBJECTIVE PROBLEMS

1. Evaluate $-9x^{n-2} - 8x^{n-1} - \left(-9x^{n-2}\right) - 8\left(x^{n-2} - 2x^{n-1}\right)$, where x = 9, n = 3.

2. Given $x^3 + 4x^2y + axy^2 + 3xy - b\,x^cy + 7xy^2 + dxy + y^2 = x^3 + y^2$ for any real numbers x and y, find the value of a, b, c, d.

3. Given that real numbers m, x, y satisfies (i) $\dfrac{2}{3}(x-5)^2 + 5m^2 = 0$; (ii) $-2a^2b^{y+1}$ and $3a^2b^3$ are like terms, find the value of the expression

   $\dfrac{3}{8}x^2y + 5m^2 - \left\{-\dfrac{7}{16}x^2y + \left[-\dfrac{1}{4}xy^2 - \dfrac{3}{16}x^2y - 3.475xy^2\right] - 6.275xy^2\right\}$.

4. Expand: $\left(5xy - 3x^2 + \dfrac{y^2}{2}\right) \times \left(5xy + 3x^2 - \dfrac{y^2}{2}\right)$

5. Given that the monomials $\dfrac{3}{4}x^b y^c$ and $-\dfrac{1}{2}x^{m-1}y^{2n-1}$ are like terms, and their sum is $\dfrac{5}{4}ax^n y^m$, find the value of abc.

6. If a, b, c are positive constants, solve the equation: $\dfrac{x-a-b}{c} + \dfrac{x-b-c}{a} + \dfrac{x-c-a}{b} = 3$

7. Solve the equation $ax + b - \dfrac{5x + 2ab}{5} = \dfrac{1}{4}$, also discuss them w.r.t different values of a and b.

8. Given that the equation a(2x + 3) + 3bx = 12x + 5 has infinitely many solutions for x. Find the values of a and b.

9. Evaluate the expression $(2+1)(2^2+1)(2^{2^2}+1)(2^{2^3}+1)\ldots(2^{2^{10}}+1)+1$.

10. Simplify the expression $(a^6 - b^6) \div (a^3 - b^3) \div (a^2 - ab + b^2)$

11. Given $x + y = \dfrac{5}{2}, x^2 + y^2 = \dfrac{13}{4}$, find the value of $x^5 + y^5$.

12. For any real numbers a, b and c, find the smallest possible values that the expression can take: $3a^2 + 27b^2 + 5c^2 - 18ab - 30c + 237$.

13. Given a - b = 2, b - c = 4, find the value of $a^2 + b^2 + c^2 - ab - bc - ca$.

14. Factorize the following:

   (i) $\left(d^2 - c^2 + a^2 - b^2\right)^2 - 4(bc - da)^2$

   (ii) $64x^6 - 729y^{12}$

   (iii) $2a^3 + 6a^2 + 6a + 18$

   (iv) $(2y - 3z)^3 + (3z - 4x)^3 + 8(2x - y)^3$

   (v) $2x^2 + 7xy - 4y^2 - 3x + 6y - 2$

   (vi) $(x + y + z)^3 - x^3 - y^3 - z^3$

   (vii) $(x^2 + x - 1)^2 + x^2 + x - 3$

   (viii) $(x - y)^3 + (y - x - 2)^3 + 8$

   (ix) $2(x^2 + 6x + 1)^2 + 5(x^2 + 1)(x^2 + 6x + 1) + 2(x^2 + 1)^2$

15. If $x^2 + 2x + 5$ is a factor of $x^4 + ax^2 + b$, find the value of $a + b$.

16. If $\dfrac{a}{b} = \dfrac{c}{d} = \dfrac{e}{f}$, then find the value of $\dfrac{2a^4b^2 + 3a^2c^2 - 5e^4f}{2b^6 + 3b^2d^2 - 5f^5}$ in terms of a and b.

17. Expand the following:

    (i) $(2a - 3b + 4c - 5d)^2$

    (ii) $(1 - x + x^2 - x^3)^2$

18. Simplify the following:

    (i) $\left(a + \dfrac{1}{a}\right)^3 - \left(a - \dfrac{1}{a}\right)^3$

    (ii) $\left(a + \dfrac{1}{a}\right)^4 - \left(a - \dfrac{1}{a}\right)^4$

19. Given that $a + \dfrac{1}{a} = 4$, find the values of $a^4 + \dfrac{1}{a^4}$ & $a^5 + \dfrac{1}{a^5}$ _____

20. if $x + y + z = 0$ then prove that $x^3 + y^3 + z^3 = 3xyz$ by using the identity $(a+b)^3 = a^3 + 3a^2b + 3ab^2 + b^3$.

21. A horse and a donkey met on their way. The donkey said to the horse: "If you transfer one bag to me, my load would have been twice of your load." The horse replied: "If you transfer one bag to me, our load would have been even." Find the number of bags on the donkey.

22. A, B, C, and D together, have 45 books. If A has 2 less, B has 2 more, C has double, and D is halved, then each would have the same number of books. How many books has A?

23. There were 140 black chocolate bars and white chocolate bars on shelve. After one quarter of the black chocolate bars was sold, the storekeeper added another 50 white chocolate bars on the shelve. Then, the number of white chocolate bars would be twice the number of black ones. Find the number of black chocolate bars at first.

24. If $n > 1$, compare $\dfrac{n}{n-1}, \dfrac{n-1}{n}, \dfrac{n}{n+1}$.

25. Given that when $x = 7$, the value of $ax^5 + bx - 8$ is 4.

    Find the value of $\dfrac{a}{2}x^5 + \dfrac{b}{2}x + 3$, when $x = 7$.

26. Given that $\dfrac{a+b}{a-b} = 7$, find the value of $\dfrac{2(a+b)}{a-b} + \dfrac{b-a}{3(a+b)}$.

27. If $x + y = 2z$, and $x \neq y$, then find the value of $\dfrac{x}{x-z} + \dfrac{y}{y-z}$.

28. If $\dfrac{1}{x} - \dfrac{1}{z} = 4$, find the value of $\dfrac{2x + 4xz - 2z}{x - z - 2xz}$.

29. Given that positive integers p, q, $p - q$ are primes, and also that $p + q$ is even evaluate the value of $\left(1 + \dfrac{1}{2}\right)^p \left(1 - \dfrac{1}{3}\right)^q$.

30. If a, b, c are rational numbers, $a + b + c = 0$ and $abc > 0$, find the value of $\dfrac{b+c}{|a|} + \dfrac{c+a}{|b|} + \dfrac{a+b}{|c|}$

31. Given that $x + y = 10$, $x^3 + y^3 = 100$, find the value of $x^2 + y^2$.

32. If $(2000 - a)(1998 - a) = 1999$, then find the value of $(2000 - a)^2 + (1998 - a)^2$.

33. For all value of k, $x = -1$ is always the solution of the equation $\dfrac{kx + a}{2} - \dfrac{2x - bk}{3} = 1$, find the value of a and of b.

34. If $abc \neq 1$, find all possible value of $\dfrac{a}{|a|} + \dfrac{b}{|b|} + \dfrac{c}{|c|}$.

35. If $(x - a)(x - 4) - 1 \equiv (x + b)(x + c)$ and a, b, c are integers, then find a.

36. If $abc = 1$, find the value of $\dfrac{a}{ab + a + 1} + \dfrac{b}{bc + b + 1} + \dfrac{c}{ac + c + 1}$.

37. Given that $\dfrac{a}{a^2 + a + 1} = 6$, find the value of $\dfrac{a^2}{a^4 + a^2 + 1}$.

38. Given that $a = 1 + \dfrac{1}{2} + \dfrac{1}{4} + \dfrac{1}{8} + \ldots + \dfrac{1}{1024}$, find the value of $\dfrac{1}{a - 1}$.

39. If $|x| = 5$, $|y| = 3$, and $|x - y| = y - x$, find the value of $(x + y)^{|x+y|}$, giving your answer in the form of a^n where a is a prime and n is a positive integer.

40. Given that the equations in x: $3\left[x - 2\left(x + \dfrac{a}{3}\right)\right] = 2x$ and $\dfrac{3x + a}{3} - \dfrac{1 + 4x}{6} = 0$ have a common solution. Find the common solution.

THANKS!

Keep smiling!

Visit Us: https://www.mathsarc.com

ANSWER KEY - Algebraic Expressions and Identities

SINGLE OPTION CORRECT

1. A	2. C	3. C	4. A
5. C	6. A	7. B	8. A
9. B	10. C	11. A	12. B
13. A	14. B	15. A	16. A
17. D	18.	19.	20.
21.	22. D	23.	24. A
25. A	26.	27. B	28.
29.	30.		

INTEGER TYPE

1. 3
2. 2
3. 0
4. 1
5. 1
6. 1
7. 2
8. 0
9. 2

SUBJECTIVE

1. 576
2. $a = -7, b = 4, c = 2, d = -3$.
3. 250
4. $-9x^4 + 28x^2y^2 - \dfrac{y^2}{4}$
5. 6/5
6. $x = a + b + c$
8. $a = 5/3, b = 26/9$.
9. 2^{2048}
10. $a + b$
11. $\dfrac{275}{32}$
12. 192
13. 28
15. 31
16. $\dfrac{a^4}{b^4}$
21. 7
22. 12
23. 76
24. $\dfrac{n}{n-1} > \dfrac{n}{n+1} > \dfrac{n-1}{n}$
25. 9
26. $13\dfrac{20}{21}$
27. 2
28. 2/3
29. 9/4
30. 1
31. 40
33. $a = 2/3, b = 3/2$.
37. $-36/11$
38. $\dfrac{1024}{1023}$

Mathsarc Education

A learning place to fulfill your dream of success!

MATHEMATICS **IIT FOUNDATION**

NUMBER SYSTEM

11:11 – A divine realm shift, Universe communication!

SINGLE OPTION CORRECT

1. Each of the numbers 1, 2, 3, and 4 is substituted, in some order for $p, q, r,$ and s. The greatest possible value of $p^q + r^s$ is

   (A) 14 (B) 162 (C) 66 (D) 83

2. The product of 20^{50} and 50^{20} is written as an integer in expanded form. The number of zeros at the end of the resulting integer is

   (A) 70 (B) 71 (C) 90 (D) 140

3. If $\dfrac{97}{19} = w + \dfrac{1}{x + \dfrac{1}{y}}$ where w, x, y are all positive integers, then $w + x + y$ equals

   (A) 16 (B) 17 (C) 18 (D) 26

4. The value of $\left(\sqrt{169} - \sqrt{25}\right)^2$ is

   (A) 64 (B) 8 (C) 16 (D) 144

5. The value of $\dfrac{5^6 \times 5^9 \times 5}{5^3}$ is

   (A) 5^{18} (B) 25^{18} (C) 5^{13} (D) 5^{51}

6. What is one half of 1.2×10^{30} ?

   (A) 6.0×10^{30} (B) 6.0×10^{29} (C) 1.2×10^{15} (D) 6.0×10^{15}

7. If 1998 = $p^s q^t r^u$, where p, q and r are prime numbers, what is the value of $p + q + r$?

    (A) 222 (B) 48 (C) 42 (D) 66

8. The numbers 123 456 789 and 999 999 999 are multiplied. How many of the digits in the final result are 9's?

    (A) 0 (B) 1 (C) 2 (D) 17

9. Which of the following numbers divide exactly into $(15 + \sqrt{49})$?

    (A) 3 (B) 4 (C) 7 (D) 11

10. If $w = 2^{129} \times 3^{81} \times 5^{128}, x = 2^{127} \times 3^{81} \times 5^{128}, y = 2^{126} \times 3^{82} \times 5^{128}$ and $z = 2^{125} \times 3^{82} \times 5^{129}$, then the order from smallest to largest is

    (A) w, x, y, z (B) x, w, y, z (C) x, y, z, w (D) z, y, x, w

11. If $y = 6 + \dfrac{1}{6}$, then $\dfrac{1}{y}$ is

    (A) 6/37 (B) 37/6 (C) 6/7 (D) 7/6

12. In the expression $\dfrac{a}{b} + \dfrac{c}{d} + \dfrac{e}{f}$ each letter is replaced by a different digit from 1, 2, 3, 4, 5, and 6. What is the largest possible value of this expression?

    (A) $8\dfrac{2}{3}$ (B) $9\dfrac{5}{6}$ (C) $9\dfrac{1}{3}$ (D) $10\dfrac{1}{3}$

13. The least value of x which makes $\dfrac{24}{x-4}$ an integer is

    (A) - 44 (B) - 28 (C) - 20 (D) - 8

14. The largest four-digit number whose digits add to 17 is 9800. The 5th largest four-digit number whose digits have a sum of 17 is

    (A) 9521 (B) 9620 (C) 9611 (D) 9602

15. Which of the following numbers is the *largest*?

    (A) 3.2571 (B) 3.2$\overline{571}$ (C) 3.2$\overline{57}$1 (D) 3.25$\overline{71}$

16. An integer x is chosen so that 3x + 1 is an even integer. Which of the following must be an odd integer?

    (A) x + 3 (B) 2x (C) 7x + 3 (D) 5x + 3

17. The value of $(256)^{0.16}(256)^{0.09}$ is

    (A) 4 (B) 16 (C) 64 (D) 256.25

18. The number of digits in $4^{16} 5^{25}$ is

    (A) 31 (B) 29 (C) 28 (D) 27

19. The people of Evenland never use odd digits. Instead of counting 1, 2, 3, 4, 5, 6, an Evenlander counts 2, 4, 6, 8, 20, 22. What is an Evenlander's version of the integer 111?

    (A) 822 (B) 842 (C) 828 (D) 824

20. The value of $0.\overline{1} + 0.\overline{12} + 0.\overline{123}$ is

    (A) $0.\overline{343}$ (B) $0.\overline{355}$ (C) $0.3\overline{5}$ (D) $0.\overline{355446}$

21. The symbol $\begin{array}{|c|c|}\hline a & b \\\hline c & d \\\hline\end{array}$ equals ad − bc. If $\begin{array}{|c|c|}\hline x-1 & 2 \\\hline 3 & -5 \\\hline\end{array} = 9$, the value of x is

    (A) -4 (B) -3 (C) -2 (D) 2

22. The value of $\sqrt{36 \times \sqrt{16}}$ is

    (A) 12 (B) 144 (C) 24 (D) 26

23. If a, b and c are positive integers with a × b = 13, b × c = 52, and c × a = 4, the value of a × b × c is

    (A) 2704 (B) 52 (C) 676 (D) 208

24. When the expression $2005^2 + 2005^0 + 2005^0 + 2005^5$ is evaluated, the final two digits are

    (A) 52 (B) 25 (C) 20 (D) 50

25. Integers m and n are each greater than 100. If m + n = 300, then m : n could be equal to

    (A) 17 : 8 (B) 5 : 3 (C) 4 : 1 (D) 3 : 2

26. Two 3 - digit numbers, abc and def, have the following property: a b c + d e f = 1000. None of a, b, c, d, e, or f is 0. What is a + b + c + d + e + f?

    (A) 10   (B) 28   (C) 21   (D) none of these

27. If $3 \leq p \leq 10$ and $12 \leq q \leq 21$, then the difference between the largest and smallest possible values of p/q is

    (A) 29/42   (B) 29/5   (C) 19/70   (D) 19/84

28. How many triples (a, b, c) of positive integers satisfy the conditions $6ab = c^2$ and $a < b < c \leq 35$?

    (A) 10   (B) 8   (C) 6   (D) 7

29. The square of an odd integer must be of the form:

    (A) 6n + 1   (B) 6n + 3

    (C) 8n + 1   (D) 4n + 1 but may not 8n + 1

30. The value of $\sqrt{(a-b)^2} + \sqrt{(b-a)^2}$ is :

    (A) Always Zero (B) Never Zero   (C) Positive iff a > b   (D) Positive only if a ≠ b

31. Which of the following is an irrational number?

    (A) $\sqrt{41616}$ (B) 23.232323

    (C) $\dfrac{(1+\sqrt{3})^3 - (1-\sqrt{3})^3}{\sqrt{3}}$ (D) 23.10100100010000.....

32. On dividing 2272 as well as 875 by a 3-digit number N, we get the same remainder in each case. The sum of the digit of N is -

    (A) 10   (B) 11   (C) 12   (D) 13

33. The smallest real number in $\{\sqrt[3]{5}, \sqrt[4]{8}, \sqrt[6]{16}, \sqrt{2}\}$

    (A) $\sqrt[3]{5}$ (B) $\sqrt[4]{8}$ (C) $\sqrt[6]{16}$ (D) $\sqrt{2}$

34. Simplest form of $\dfrac{(20)^{31} \times (12)^5 \times (30)^3}{(10)^{13} \times (6)^7}$ is

   (A) $2^{52} \times 3 \times 5^{21}$
   (B) $2^{55} \times 3^3 \times 5^{21}$
   (C) $2^{55} \times 3 \times 5^{21}$
   (D) $2^{55} \times 3 \times 5^{31}$

35. The expression $\sqrt{12+6\sqrt{3}} + \sqrt{12-6\sqrt{3}}$ simplifies to

   (A) 4
   (B) $2\sqrt{3}$
   (C) $3\sqrt{3}$
   (D) 6

36. If $m = 3^{n-1}$ & $3^{4n-1} = 27$, what is the value of m + n ?

   (A) 3
   (B) 2
   (C) 0
   (D) 1

37. Select the correct option

   (A) $\sqrt[3]{4} > \sqrt[4]{5} > \sqrt[4]{3}$
   (B) $\sqrt[4]{5} > \sqrt[3]{4} > \sqrt[4]{3}$
   (C) $\sqrt[4]{5} < \sqrt[3]{4} < \sqrt[3]{3}$
   (D) None of these

38. If $\sqrt{2} - \sqrt{3} + \sqrt{k}$ is rational, then k =

   (A) $\sqrt{3} - \sqrt{2}$
   (B) $5 - 2\sqrt{6}$
   (C) $-5 - 2\sqrt{6}$
   (D) 5

39. If P is an integer and P^2 is divisible by 3, then P is divisible by 3. This statement is

   (A) Always true
   (B) never true
   (C) true when P is positive
   (D) true when P is negative

40. $\sqrt[6]{8a^3b} \times \sqrt[3]{4a^2b^3}$ =

   (A) $2ab\sqrt[6]{2ab}$
   (B) $2a^3b\sqrt[6]{2ab}$
   (C) $4ab\sqrt[6]{2ab}$
   (D) $4ab\sqrt[6]{4ab}$

41. The value of $\dfrac{2^{m+3} \times 2^{2m-n} \times 5^{m+n+3} \times 6^{n+1}}{6^{m+1} \times 10^{n+3} \times 15^m}$ is equal to

   (A) 0
   (B) 1
   (C) 2^m
   (D) None of these

42. If $\dfrac{\sqrt{2+x} + \sqrt{2-x}}{\sqrt{2+x} - \sqrt{2-x}} = 2$, then x is

   (A) 2/3
   (B) 3/4
   (C) 8/5
   (D) 1/7

43. If $\dfrac{x}{3} = \dfrac{y}{1} = \dfrac{z}{2}$ and $2x - 3y + 4z = 10$, then $11(x + y + z) =$

   (A) 20   (B) 40   (C) 60   (D) 80

44. $5\sqrt[3]{250} + 7\sqrt[3]{16} - 14\sqrt[3]{54} =$

   (A) $2\sqrt[3]{2}$ (B) $-3\sqrt[3]{2}$ (C) $13\sqrt{6}$ (D) None of these

45. $\dfrac{(25)^{3/2} \times (243)^{3/5}}{(16)^{5/4} \times (8)^{4/3}} =$

   (A) $\dfrac{216}{17}$ (B) $\dfrac{3375}{12}$ (C) $\dfrac{5733}{17}$ (D) None of these

MULTIPLE OPTIONS CORRECT

1. If $x = 3$, which of the following expressions is an even number(s)?

   (A) $9x + 3$ (B) x^3 (C) $2(x^2 + 9)$ (D) $3x^2$

2. Select the correct statements

   (A) Decimal expansion of $\sqrt{3}$ is non-terminating and recursive

   (B) Decimal expansion of $\sqrt{2}$ is non-terminating and non-recursive

   (C) Rationalizing factor of $5 + \sqrt{3}$ is $5 - \sqrt{3}$

   (D) The decimal expansion of $\dfrac{1387657}{2^{32} \times 5^{41}}$ will terminate

3. If $\dfrac{2+\sqrt{3}}{3-\sqrt{3}} = a + b\sqrt{3}$ where $a, b \in Q$ then

   (A) $a > b$ (B) $a + b = 7/3$ (C) $a/b = 5/9$ (D) $a - b = 2/3$

4. If $\dfrac{\sqrt{7}-1}{\sqrt{7}+1} - \dfrac{\sqrt{7}+1}{\sqrt{7}-1} = a + b\sqrt{7}$, then

   (A) $a > b$ (B) $b^a = 1$ (C) $b/a = 0$ (D) $|b| > a$

5. Select the correct options

   (A) $\sqrt{2}-1=\sqrt{3-2\sqrt{2}}$

   (B) $2\sqrt[4]{2} > \sqrt{5}$

   (C) $2\sqrt[5]{3} > 3\sqrt{7}$

   (D) $\sqrt{4} = \sqrt{(-2)(-2)} = \sqrt{2 \times 2} = \pm 2$

6. Select the correct statements (m, n are two Natural Numbers)

   (A) 143 is a prime number (B) $\pi = \dfrac{22}{7}$ (C) $\sqrt{3}$ is an irrational number

   (D) $\dfrac{1}{2^m \cdot 5^n}$ is terminating rational number

7. 120^3 can be written as

   (A) $(2^3)^3 \cdot 27 \cdot (5)^3$
   (B) $(2^3)^3 \cdot 3^3 \cdot (5)^3$
   (C) $(40)^3 \cdot (3)^3$
   (D) $2^{27} \cdot (3)^3 \cdot (5)^3$

8. If $N = \sqrt{3 - 2\sqrt{2}}$, then

   (A) $N - \sqrt{2}$ is an irrational number

   (B) $N - \sqrt{2}$ is a rational number

   (C) $N - \sqrt{3}$ is a rational number

   (D) if $N = p + q\sqrt{r}$, where p, q, r are integers, Then p + q + r = 2.

9. Select the correct statements

   (A) N = {1, 2, 3, 4,........., ∞}
   (B) W = {0, 1, 2, 3, 4,......., ∞}

   (C) Z⁺ = {1, 2, 3, 4,........., ∞}
   (D) Q = R - Qᶜ

10. Select the correct options

    (A) $\left(\left((625)^{-1/2}\right)^{-1/4}\right)^2 = 5$

    (B) $\left(5\left(8^{1/3} + 27^{1/3}\right)^3\right)^{1/4} = 5$

    (C) $\left(\sqrt[4]{\left(\dfrac{1}{x}\right)^{-12}}\right)^{-2/3} = \dfrac{1}{x}$

    (D) $\dfrac{\sqrt{x^3} \times \sqrt[3]{x^5}}{\sqrt[5]{x^3}} \times \sqrt[30]{x^{77}} = x^{77/15}$

INTEGER TYPE

1. If $x = \dfrac{1}{\sqrt{2-\sqrt{3}}}$ and $y = \dfrac{1}{\sqrt{2+\sqrt{3}}}$ then the value of $(x-y)^2$ is ____

2. If $2^{5x} \div 2^x = \sqrt[5]{2^{20}}$ then x equals?

3. The sum of the digits of the number $2^{2000} \, 5^{2002}$ is ____

4. If $a^m \, a^n = a^{mn}$, then value of $m(n-2) + n(m-2)$ is ____

5. If a, b, c are three positive real number and $x \neq 0, x \neq 1$, then $\left(\dfrac{x^a}{x^b}\right)^{1/ab} \times \left(\dfrac{x^b}{x^c}\right)^{1/bc} \times \left(\dfrac{x^c}{x^a}\right)^{1/ac}$ equals ____

6. The value of $(2-\sqrt{3}) \times \dfrac{(\sqrt{3}+1)}{(\sqrt{3}-1)}$ is ____

7. If $5^{3x-1} \cdot 3^x = 75$ then x is ____

8. Express $\left(\left(2^{1/2} \cdot 4^{3/4} \cdot 8^{5/6} \cdot 16^{7/8} \cdot 32^{9/10}\right)^4\right)^{1/25}$ as an integer.

SUBJECTIVE PROBLEMS

1. Simplify each of the following

   (i) $\left(a^{2/3}\right)^3 \div a^{3/2} \times a^{-3}$

   (ii) $\left(c^{-2}\right)^{2/3} \times c^{3/2} \div c^{-3}$

   (iii) $\left(\left(a^{-3}\right)^{4/7} \times \left(a^{9/2}\right)\right)^{3/2}$

   (iv) $\left[\left(\dfrac{2}{x}\right)^{-3} \times \left(\dfrac{3}{x}\right)^4\right]^{3/4}$

   (v) $\left(5t^2\right)^2 \div (25t)^{3/2} \times \left(5t^{1/2}\right)^3$

   (vi) $\left\{\dfrac{a^{p-q}}{\sqrt[q]{a^{q^2-pq}}} \times a^{2(p-q)}\right\}^n$

   (vii) $\dfrac{3 \cdot 2^n - 4 \cdot 2^n}{2^n - 2^{n-1}}$

   (viii) $\sqrt[ab]{\dfrac{x^a}{x^b}} \times \sqrt[bc]{\dfrac{x^b}{x^c}} \times \sqrt[ca]{\dfrac{x^c}{x^a}}$

(ix) $\sqrt[3]{x^4 y \times \dfrac{1}{4x^4 y^3}}$

(x) $\dfrac{a^{1/2}+a^{-1/2}}{1-a}+\dfrac{1-a^{-1/2}}{1+\sqrt{a}}$

2. Solve for (x, y): $x^{y-2} = 4$ and $x^{2y-3} = 64$.

3. If $4^x = 5^y = 20^z$, then show that $z = \dfrac{xy}{x+y}$.

4. If $(p+q)^{-1}(P^{-1}+q^{-1}) = p^a q^b$, then prove that $a + b + 2 = 0$.

5. Simplify: $\dfrac{\left(x^{2^{n-1}}+y^{2^{n-1}}\right)\left(x^{2^{n-1}}-y^{2^{n-1}}\right)}{x^{2^n}-y^{2^n}}$.

6. If $A = \dfrac{1}{1+x^{b-c}+x^{b-a}} + \dfrac{1}{1+x^{c-a}+x^{c-b}} + \dfrac{1}{1+x^{a-b}+x^{a-c}}$

   and $B = \left(\dfrac{x^m}{x^n}\right)^{m^2+mn+n^2} \times \left(\dfrac{x^n}{x^l}\right)^{n^2+nl+l^2} \times \left(\dfrac{x^l}{x^m}\right)^{l^2+lm+m^2}$. Find the simplest values A and B.

7. Arrange the following in ascending order: 2^{5555}, 3^{3333}, 6^{2222}

8. (a) Find all integers n such that $(n^2 - n - 1)^{n+2} = 1$.

   (b) If $x = \dfrac{4ab}{a+b}$, find the value of $\dfrac{x+2a}{x-2a}+\dfrac{x+2b}{x-2b}$

9. (a) Find all the positive perfect cubes that divide 9^9.

   (b) Find the integer closest to $100(12 - \sqrt{143})$.

   (c) If $8^{2x} = 16^{1-2x}$, find the value of 3^{7x}.

   (d) If $\dfrac{4^x}{2^{x+y}} = 8$ and $\dfrac{9^{x+y}}{3^{5y}} = 243$ find the value of x - y.

10. Simplify:

    (i) $5\sqrt{45}+\sqrt{20}$

    (ii) $\sqrt{27}+3\sqrt{75}-2\sqrt{12}$

    (iii) $5\sqrt{24}-4\sqrt{32}+3\sqrt{18}-2\sqrt{54}$

    (iv) $2\times\sqrt{2}\times\sqrt[4]{2}\times\sqrt[8]{2}$

    (v) $\sqrt{240}+\sqrt{40}$

11. Express as roots of rational numbers:

    (i) $4\sqrt{20} \times \sqrt{80}$

    (ii) $\left(5\sqrt{3} + 4\sqrt{2}\right) \times \left(3\sqrt{3} - 2\sqrt{2}\right)$

    (iii) $2^{3/4} \times 3^{-1/2} \times 4^{1/8} \times 9^{4/3}$

    (iv) $\sqrt[7]{\left(5^2 \times 3^{14}\right)} \times \sqrt[14]{15^3}$

12. Rationalize the denominator of:

    (i) $\dfrac{1}{\sqrt{5} - \sqrt{30}}$

    (ii) $\dfrac{3\sqrt{2} - 2\sqrt{5}}{3\sqrt{5} - 4\sqrt{2}}$

    (iii) $\dfrac{1}{1 + \sqrt{2} - \sqrt{3}}$

    (iv) $\dfrac{\sqrt{2}}{\sqrt{2} + \sqrt{3} - \sqrt{5}}$

13. Simplify: $T = \dfrac{1}{3 - \sqrt{8}} - \dfrac{1}{\sqrt{8} - \sqrt{7}} + \dfrac{1}{\sqrt{7} - \sqrt{6}} - \dfrac{1}{\sqrt{6} - \sqrt{5}} + \dfrac{1}{\sqrt{5} - 2}$.

14. Simplify: $2 + \sqrt{2} + \dfrac{1}{2 + \sqrt{2}} + \dfrac{1}{\sqrt{2} - 2}$

15. Simplify: $\left(\sqrt[3]{\sqrt{2014} + 1} - \sqrt[3]{\sqrt{2014} - 1} + \sqrt[6]{2013}\right)^3 + \left(\sqrt[3]{\sqrt{2014} + 1} - \sqrt[3]{\sqrt{2014} - 1} - \sqrt[6]{2013}\right)^3$.

ANSWER KEY – Number System

SINGLE OPTION CORRECT

1. D	2. C	3. A	4. A
5. C	6. B	7. C	8. A
9. D	10. C	11. A	12. B
13. C	14. C	15. D	16. D
17. A	18. C	19. B	20. D
21. C	22. A	23. B	24. A
25. D	26. B	27. A	28. B
29. C	30. D	31. D	32. A
33. D	34. C	35. D	36. B
37. A	38. B	39. A	40. A
41. D	42. C	43. C	44. B
45. D			

MULTI OPTIONS CORRECT

1. A, C	2. B, C, D	3. A, B, D	4. A, B, D
5. A, B	6. C, D	7. A, B, C	8. B, D
9. A, B, C, D	10. A, B, D		

INTEGER TYPE

1. 2	2. 1	3. 7	4. 0
5. 1	6. 1	7. 1	8. 4

SUBJECTIVE

2. $(x, y) = (4, 3)$ 6. $A = B = 1$

Mathsarc Education

A learning place to fulfill your dream of success!

MATHEMATICS IIT FOUNDATION

NUMBER SYSTEM - NTSE

11:11 – A divine realm shift, Universe communication!

SINGLE OPTION CORRECT

1. The value of the expression $\dfrac{1}{\sqrt{11-2\sqrt{30}}} - \dfrac{3}{\sqrt{7-2\sqrt{10}}} - \dfrac{4}{\sqrt{8+4\sqrt{3}}}$ after simplification is

   (A) $\sqrt{30}$ (B) $2\sqrt{10}$ (C) 1   (D) 0

2. A rational number between $\sqrt{2}$ and $\sqrt{3}$ is:

   (A) 1.5   (B) $\dfrac{\sqrt{2}+\sqrt{3}}{2}$ (C) $\sqrt{2} \times \sqrt{3}$ (D) 1.8

3. If n is a natural number, then which number always ends at 6 from the following?

   (A) 4^n (B) 2^n (C) 6^n (D) 8^n

4. Expressing $0.\overline{34} + 0.3\overline{4}$ as a single decimal, we get

   (A) $0.67\overline{88}$ (B) $0.6\overline{89}$ (C) $0.6\overline{878}$ (D) $0.6\overline{87}$

5. Given that $\dfrac{1}{7} = 0.\overline{142857}$ which is a repeating decimal having six different digits. If x is the sum of such first three positive integers n such that $\dfrac{1}{n} = 0.\overline{abcdef}$, where a, b, c, d, e and f are different digits, then the value of x is

   (A) 20   (B) 21   (C) 41   (D) 42

6. Which of the following digits is ruled out in the unit's place of $12^n + 1$ for every positive integer n?

   (A) 1   (B) 3   (C) 5   (D) 7

7. The value of $\sqrt{97 \times 98 \times 99 \times 100 + 1}$ is equal to

   (A) 9901 (B) 9891 (C) 9801 (D) 9701

8. Which of the following is a square root of $21 - 4\sqrt{5} + 8\sqrt{3} - 4\sqrt{15}$.

   (A) $2\sqrt{3} - 2 - \sqrt{5}$ (B) $\sqrt{5} - 3 + 2\sqrt{3}$ (C) $2\sqrt{3} - 2 + \sqrt{5}$ (D) $2\sqrt{3} + 2 - \sqrt{5}$

9. If a and b are positive numbers and c and d are real numbers, positive or negative, then $a^c \le b^d$.

   (A) if $a \le b$ and $c \le d$

   (B) if either $a \le b$ or $c \le d$

   (C) if $a \ge 1, b \ge 1, d \ge c$

   (D) is not implied by any of the foregoing conditions

10. Simplify: $\left(\sqrt[3]{\sqrt[6]{a^9}}\right)^4 \left(\sqrt[6]{\sqrt[3]{a^9}}\right)^4$

    (A) a^{16} (B) a^{12} (C) a^8 (D) a^4

11. The unit digit in the expression $55^{725} + 73^{5810} + 22^{853}$ is -

    (A) 0 (B) 4 (C) 5 (D) 6

12. If $x + y + z = 0$ & $x \ne 0, y \ne 0, z \ne 0$ then the value of $\dfrac{x^2}{yz} + \dfrac{y^2}{xz} + \dfrac{z^2}{xy}$ is

    (A) 0 (B) 1 (C) 2 (D) 3

13. If $(2^x - 4)^3 + (4^x - 2)^3 = (4^x + 2^x - 6)^3$, then the sum of all real values of x is

    (A) 0.5 (B) 1.5 (C) 2.5 (D) 3.5

14. If $2019^x + 2019^{-x} = 3$, then the value of $\sqrt{\dfrac{2019^{6x} - 2019^{-6x}}{2019^x - 2019^{-x}}}$ is:

    (A) 3 (B) 6 (C) 9 (D) 12

15. If $\dfrac{1}{x+y} = \dfrac{1}{x} + \dfrac{1}{y}$, then the value of $\left(\dfrac{x}{y}\right)^6 + \left(\dfrac{x}{y}\right)^3$ is

   (A) 0 (B) 1/2 (C) 1 (D) 2

16. If $\sqrt{p} - \sqrt{q} = 20$, then the maximum value of $\left(\dfrac{p-5q}{100}\right)$ is:

   (A) 5 (B) 10 (C) 15 (D) 25

17. If $N = \sqrt[3]{4} + \sqrt[3]{2} + 1$, then the value of $\dfrac{1}{N^3} + \dfrac{3}{N^2} + \dfrac{3}{N}$ is:

   (A) 2 (B) 4 (C) 7 (D) 1

18. In the decimal expansion of a rational number $\dfrac{14580}{625 \times 3}$, there are ____ digits (nos) after decimal.

   (A) 2 (B) 3 (C) 4 (D) 5

19. If $\dfrac{3}{\sqrt{28+10\sqrt{3}} - \sqrt{7-4\sqrt{3}}} = a + \sqrt{3}b$, where a and b are integers, then value of $\sqrt{5a+12b}$ is:

   (A) 4 (B) 3 (C) √11 (D) √13

20. If a, b and c are integers such that $(\sqrt[3]{4} + \sqrt[3]{2} - 2)(\sqrt[3]{4a} + \sqrt[3]{2b} + c) = 20$, then which one of the following is true?

   (A) a + b − c = 10 (B) a − b + c = 10 (C) a + b = 2c (D) a + b + c = 16

21. Sum of the first n terms of the series $\sqrt{2} + \sqrt{8} + \sqrt{18} + \ldots$ is

   (A) $\dfrac{n(n+1)}{2}$ (B) $\sqrt{2}n$ (C) $\dfrac{n(n+1)}{\sqrt{2}}$ (D) 1

22. If $x = \dfrac{y}{y+1}$ and $y = \dfrac{a-2}{2}$, then the value of $x(y+2) + \dfrac{x}{y} + \dfrac{y}{x}$ is

   (A) 1 (B) 0 (C) -1 (D) a

23. If $x^2 + y^2 = 2\sqrt{2}x + 4\sqrt{2}y - 10$, then the value of $\dfrac{x}{y}$ is

   (A) 1/2 (B) 1/4 (C) 2 (D) 4

24. $\sqrt{a\sqrt{b\sqrt{c\sqrt{d}}}} =$

   (A) $a^{1/2}b^{1/4}c^{1/8}d^{1/16}$ (B) $(abcd)^{1/16}$ (C) $(abcd)^{1/8}$ (D) $a^{1/2}b^{1/2}c^{1/2}d^{1/2}$

INTEGER TYPE

1. Simplify: $\dfrac{\left(5\sqrt{3}+\sqrt{50}\right)\left(5-\sqrt{24}\right)}{\left(\sqrt{75}-5\sqrt{2}\right)}$

2. Simplify: $\dfrac{3\sqrt{2}}{\sqrt{6}+\sqrt{3}} - \dfrac{4\sqrt{3}}{\sqrt{6}+\sqrt{2}} + \dfrac{\sqrt{6}}{\sqrt{3}+\sqrt{2}}$

3. Simplify: $\dfrac{\sqrt{\left(6+2\sqrt{3}+2\sqrt{2}+2\sqrt{6}\right)}-1}{\sqrt{5+2\sqrt{6}}}$

4. Find the value of $\dfrac{2\cdot 3^{n+1} - 7\cdot 3^{n-1}}{3^{n+1} + 2\left(\dfrac{1}{3}\right)^{1-n}}$

SUBJECTIVE PROBLEMS

1. The greatest number among $\sqrt[3]{9}, \sqrt[4]{11}, \sqrt[6]{17}$ is

2. Find the value of $\left(\dfrac{1}{3}\right)^{-10} \times 27^{-3} + \left(\dfrac{1}{5}\right)^{-4} \times (25)^{-2} + \left(64^{1/9}\right)^{-3}$

3. If $\dfrac{\left(2^{n+1}\right)^m \left(2^{2n}\right)2^n}{\left(2^{m+1}\right)^n \left(2^{2m}\right)} = 1$, then find the value of m

4. Simplify: $a\left(\dfrac{\sqrt{a}+\sqrt{b}}{2b\sqrt{a}}\right)^{-1} + b\left(\dfrac{\sqrt{a}+\sqrt{b}}{2a\sqrt{b}}\right)^{-1}$

THANKS!

Keep smiling!

Visit Us: https://www.mathsarc.com

ANSWER KEY – Number System NTSE

SINGLE OPTION CORRECT

1. D   2. A   3. C   4. D
5.   6. A   7. D   8. D
9. D   10. D   11.   12. D
13.   14.   15.   16.
17.   18.   19.   20.
21.   22.   23.   24.

INTEGER TYPE

1. 1   2. 0   3. 1   4. 1

SUBJECTIVE

1. $\sqrt[3]{9}$ 2. 17/4   3. m = 2n   4. 2ab

Mathsarc Education

A learning place to fulfill your dream of success!

MATHEMATICS   **IIT FOUNDATION**

INTRODUCTION TO TRIGONOMETRY - X

TRIGONOMETRY IS A SINE OF THE TIME!

SECTION – A (Brain Nerves Development)

The section contains (20 Single Correct) chapter problems based on the concepts taught in class & asked in competitive exams. First do each & every problems of this section your own without taking help of your teacher or friends. If you find difficulties to solve & have devoted atleast 15 - 20 Minutes then discuss it with your friend and then your teachers. Marking (+3, - 1)

1. In $\triangle ABC$, If $3A = 4B = 6C$ then $A : B : C$ be –

    (A) $3 : 4 : 6$ (B) $\dfrac{1}{4} : \dfrac{1}{3} : \dfrac{1}{2}$ (C) $6 : 4 : 3$ (D) $4 : 3 : 2$

2. In $\triangle ABC$, if $\angle B = 90°$, $AB = 5$, $BC = 12$, then $\sin C = $ ....

    (A) $\dfrac{12}{13}$ (B) $\dfrac{5}{13}$ (C) $\dfrac{5}{12}$ (D) $\dfrac{13}{5}$

3. $(\sec\theta + \tan\theta)(1 - \sin\theta) = $ ....

    (A) 0 (B) 1 (C) $\cos\theta$ (D) $\sin\theta$

4. If $\tan\theta = \dfrac{1}{\sqrt{3}}$, then the value of $\dfrac{\cosec^2\theta - \sec^2\theta}{\cosec^2\theta + \sec^2\theta}$ is

    (A) $\sqrt{3}$ (B) $\dfrac{1}{3}$ (C) $\dfrac{1}{2}$ (D) $\dfrac{1}{\sqrt{3}}$

5. If the measures of the angles $\triangle ABC$ are in proportion $1 : 2 : 3$, then the measure of the smallest angle is ___

    (A) $30°$ (B) $60°$ (C) $90°$ (D) $120°$

6. In △ABC, if $\frac{AB}{1} = \frac{AC}{2} = \frac{BC}{\sqrt{3}}$, then m∠C = _____

   (A) 90°   (B) 30°   (C) 60°   (D) 45°

7. The sum of all interior angles and one exterior angle of a convex K-sided polygon is 1350°, the value of K is

   (A) 7   (B) 8   (C) 9   (D) 11

8. If cos9α = sinα and 9α < 90°, then the value of tan5α?

   (A) $\frac{1}{\sqrt{3}}$ (B) $\sqrt{3}$ (C) 1   (D) 0

9. $\sin x = \frac{6\sin 30° - 8\cos 60° + 2\tan 45°}{2(\sin^2 30° + \cos^2 60°)}$, then x =

   (A) 30°   (B) 45°   (C) 60°   (D) 90°

10. The value of $\frac{\sin 18°}{\cos 72°}$ will be

    (A) 2   (B) 3   (C) 4   (D) 1

11. The value of $\frac{\cot 54°}{\tan 36°} + \frac{\tan 20°}{\cot 70°} - 2$ is

    (A) 2   (B) 1   (C) -1   (D) 0

12. The arc length of the sector of a circle of radius R which makes an angle x° at the centre is

    (A) $\frac{\pi Rx}{180}$ (B) $\frac{\pi Rx}{90}$ (C) $\frac{\pi R^2 x}{180}$ (D) $\frac{\pi R^2 x}{360}$

13. The smallest positive solution of the equation $81^{\sin^2 x} + 81^{\cos^2 x} = 30$ is

    (A) $\frac{\pi}{12}$ (B) $\frac{\pi}{8}$ (C) $\frac{\pi}{3}$ (D) $\frac{\pi}{6}$

14. $\cos\theta\sqrt{\sec^2\theta - 1}$ is equal to

    (A) cotθ   (B) sinθ   (C) secθ   (D) 1

15. The value of tan1° tan2° tan3° ………tan89° is :

    (A) 0 (B) 1 (C) 2 (D) none of these

16. If cosec x − cot x = $\frac{1}{3}$, where x ≠ 0, then the value of $\cos^2 x - \sin^2 x$ is

    (A) $\frac{16}{25}$ (B) $\frac{9}{25}$ (C) $\frac{7}{25}$ (D) $\frac{8}{25}$

17. In the figure, ABCD is a square of side 1dm and ∠PAQ = 45°. The perimeter (in dm) of the triangle PQC is

    (A) 2 (B) 1 + √2

    (C) 2√2 - 1 (D) 1 + √3

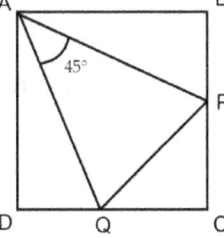

18. In the figure, ABC is a triangle in which AD bisect ∠A, AC = BC, ∠B = 72° and CD = 1cm. Length of BD (in cm) is

    (A) 1 (B) $\frac{1}{2}$

    (C) $\frac{\sqrt{5}-1}{2}$ (D) $\frac{\sqrt{3}+1}{2}$

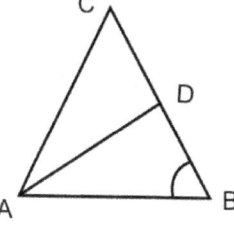

19. The value of cot 12° cot 38° cot 52° cot 60° cot 78° is:

    (A) 1 (B) 0 (C) 1/√2 (D) 1/√3

20. If θ is an acute angle such that tanθ = 2/3, then value of $\left(\frac{1+\tan\theta}{\sin\theta+\cos\theta}\right)\left(\frac{1-\cot\theta}{\sec\theta+\csc\theta}\right)$ is:

    (A) -1/5 (B) $-\frac{4}{\sqrt{13}}$ (C) 1/5 (D) $\frac{4}{\sqrt{13}}$

SECTION – B (Rank Booster Problems)

The section contains 10 single correct questions of Higher Order Thinking. The section will generate mastery in the topic. So, solve and become unique among your friends. Marking (+4, -1)

21. If $\sin\theta + \cos\theta = a$ and $\dfrac{\sin\theta + \cos\theta}{\sin\theta \cdot \cos\theta} = b$, then

    (A) $b = \dfrac{2a}{a^2 - 1}$
    (B) $a = \dfrac{2b}{b^2 - 1}$
    (C) $ab = b^2 - 1$
    (D) $a + b = 1$

22. If $\csc\theta + \cot\theta = \dfrac{11}{2}$, then $\tan\theta$ is

    (A) $\dfrac{21}{22}$
    (B) $\dfrac{15}{16}$
    (C) $\dfrac{44}{117}$
    (D) None of these

23. The value of $\tan 225° \cdot \cot 405° + \tan 765° \cdot \cot 675°$ is

    (A) 1
    (B) 0
    (C) 2
    (D) -1

24. One of the given statements is not true in general. Which one is it?

    1: $\sin^2 A + \cos^2 A = 1$
    2: $\cot A (\cot A + \tan A) = \csc^2 A$
    3: $1 + \tan^2 A = \sec^2 A$
    4: $\tan A - \cot A = \sec A - \csc A$

    (A) 1
    (B) 2
    (C) 3
    (D) 4

25. The value of $\cot A + \tan(180° + A) + \tan(90° + A) + \tan(360° - A) =$

    (A) 0
    (B) 1
    (C) -1
    (D) 2

26. If $\cos(1°)\cos(2°)\cos(3°)\cdots\cos(178°)\cos(179°) = x + 1$ then value of x is?

    (A) 0
    (B) 1
    (C) -1
    (D) none of these

27. If $\tan\theta = -\dfrac{4}{3}$, then $\sin\theta$ is equal to

    (A) $-\dfrac{4}{5}$ but not $\dfrac{4}{5}$
    (B) $-\dfrac{4}{5}$ or $\dfrac{4}{5}$
    (C) $\dfrac{4}{5}$ but not $-\dfrac{4}{5}$
    (D) None of these

28. $\dfrac{\sin(-660°)\tan(1050°)\sec(-420°)}{\cos(225°)\operatorname{cosec}(315°)\cos(510°)} =$

(A) $\dfrac{\sqrt{3}}{4}$
(B) $\dfrac{\sqrt{3}}{2}$
(C) $\dfrac{2}{\sqrt{3}}$
(D) $\dfrac{4}{\sqrt{3}}$

29. The value of $\left(1+\cos\dfrac{\pi}{8}\right)\left(1+\cos\dfrac{3\pi}{8}\right)\left(1+\cos\dfrac{5\pi}{8}\right)\left(1+\cos\dfrac{7\pi}{8}\right)$ is equal to

(A) $\dfrac{1}{4}$
(B) $\dfrac{1}{2}$
(C) $\dfrac{1}{8}$
(D) $\dfrac{1}{16}$

30. The value of $3\left[\sin^4\left(\dfrac{3\pi}{2}-\alpha\right)+\sin^4(3\pi+\alpha)\right]-2\left[\sin^6\left(\dfrac{\pi}{2}+\alpha\right)+\sin^6(5\pi-\alpha)\right]$ is:

(A) 0
(B) 1
(C) 3
(D) $\sin 4\alpha + \sin 6\alpha$

SECTION – C (In-depth Analysis, Brain Storming Problems)

The section contains 10 Multi option correct questions. The section required in-depth knowledge to answer the questions correctly. So, solve them and get a wizard level brain. Marking (+4, 0)

31. If $\sqrt{\dfrac{1-\sin A}{1+\sin A}} = K_1 \sec A - K_2 \tan A$

(A) $K_1 - K_2 = 2$
(B) $K_1 \neq K_2$
(C) $K_1 = K_2 = 1$
(D) $K_1 = K_2$

32. If $\tan A = \dfrac{3}{4}$

(A) $\dfrac{1-\cos A}{1+\cos A} = \dfrac{1}{9}$
(B) $\operatorname{cosec} A = \pm\dfrac{5}{3}$

(C) $\operatorname{cosec} A = \dfrac{5}{3}$
(D) $\sec A = \pm\sqrt{1+\left(\dfrac{3}{4}\right)^2}$

33. If $\cos\theta = -\dfrac{1}{2}$ then θ is equal to

    (A) 60° (B) 120° (C) 210° (D) 240°

34. Select the correct statements

    (A) $1 - \sin^2 5 = \cos^2 5$
    (B) $\sec^2 x - \tan^2 x = 1$
    (C) $\dfrac{1-\sin\theta}{\cos\theta} = \dfrac{1}{\sec\theta + \tan\theta}$
    (D) $\sin(a+b) = \sin a \cos b + \cos a \sin b$

35. Select the correct options

    (A) $3° = \left(\dfrac{\pi}{60}\right)^c$
    (B) $1.25° = 1°15'$
    (C) $\dfrac{\pi}{18} = 10°$
    (D) $0.125° = 7'30''$

36. If A and B are acute angles such that sinA = sin²B, 2 cos²A = 3 cos² B then

    (A) A = π/6 (B) A = π/2 (C) B = π/4 (D) B = π/3

37. Select the correct options

    (A) $\sin(30°) = \dfrac{1}{2}$
    (B) $\sin(60°) = \dfrac{\sqrt{3}}{2}$
    (C) $\cosec(45°) = 2$
    (D) $\sin^2 1° + \sin^2 89° = 1$

38. If $\sin\theta = -\dfrac{3}{5}$ & $\cos\theta = \dfrac{4}{5}$, then

    (A) $\tan\theta = \dfrac{3}{4}$
    (B) $\tan\theta = -\dfrac{3}{4}$
    (C) $\tan(2\theta) < 0$
    (D) $\cos(2\theta) = \dfrac{7}{25}$

39. If tan1° tan2° tan3° ………..tan89° = x² – 8, then the value of x can be

    (A) - 1 (B) 1 (C) - 3 (D) 3

40. If sinθ (1 + sinθ) + cosθ (1 + cosθ) = x and sinθ (1 - sinθ) + cosθ (1 - cosθ) = y then…

    (Hint: sin2θ = 2sinθ cosθ)

    (A) x² – 2x = sin2θ
    (B) y² + 2x = sin2θ
    (C) xy = sin2θ
    (D) x – y = 2

SECTION – D (INTEGER ANSWER TYPE)

This section contains 10 questions. Each question, when worked out will result in one integer from 0 to 9. (Both inclusive) Marking (+ 4, 0)

41. The value of $\left(\sec^2\theta - \dfrac{\sec^2\theta}{\csc^2\theta} + 1\right)\left(\dfrac{\sin\theta}{\csc\theta} + \dfrac{\cos\theta}{\sec\theta}\right)$ is

42. If $\cos^6\theta + \sin^6\theta = 1 + K\sin^2\theta \cdot \cos^2\theta$ then the absolute value of K is

43. The value of $6(\sin^6\theta + \cos^6\theta) - 9(\sin^4\theta + \cos^4\theta) + 4$ equals to

44. If $\sin\theta_1 + \sin\theta_2 + \sin\theta_3 = 3$, then $\cos\theta_1 + \cos\theta_2 + \cos\theta_3$ is equal to

45. The value of $\dfrac{\tan 70° - \tan 20°}{\tan 50°}$ is _____

46. The value of $\dfrac{1}{\cos\theta + \sin\theta}\left(\dfrac{\sin\theta}{1 - \cot\theta} + \dfrac{\cos\theta}{1 - \tan\theta}\right)$ is equal to

47. The value of $\sin^2\left(7\dfrac{1}{2}\right) + \cos^2\left(7\dfrac{1}{2}\right) - \left(\sin^2(30°) + \cos^2(30°)\right) + \left(\sin^2(7) + \sin^2(83)\right)$ is = ___

48. The value of $\cot A + \tan(180° + A) + \tan(90° + A) + \tan(360° - A) = ?$

49. The value of $\dfrac{\cos(90° + \theta°)\sec(-\theta°)\tan(180° - \theta°)}{\sec(360° - \theta°)\sin(180° + \theta°)\cot(\theta° - 90°)}$ is

50. If $\sin x + \sin^2 x = 1$, then the value of $\cos^{12} x + 3\cos^{10} x + 3\cos^8 x + \cos^6 x$ is

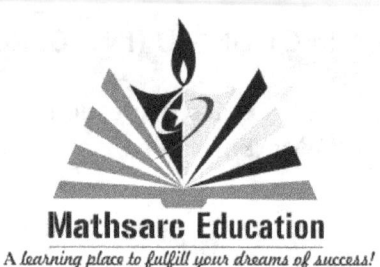

THANKS!

Keep smiling!

Visit Us: https://www.mathsarc.com

ANSWER KEY – Introduction to Trigonometry

1	D	11	D	21	A	31	A, B	41	2
2	B	12	A	22	C	32	A, B, D	42	3
3	C	13	D	23	B	33	B, D	43	1
4	C	14	B	24	D	34	A, B, C, D	44	0
5	A	15	B	25	A	35	A, B, C, D	45	2
6	B	16	C	26	C	36	A, C	46	1
7	C	17	A	27	B	37	A, B, D	47	1
8	C	18	C	28	C	38	B, C, D	48	0
9	D	19	D	29	C	39	C, D	49	1
10	D	20	A	30	B	40	A, B, C, D	50	1

Mathsarc Education

A learning place to fulfill your dream of success!

MATHEMATICS IIT JEE Main/Advanced

MODULUS

ABSOLUTE VALUE OF X IS DENOTED BY |X|.

PROPERTIES OF |X|

(i) $\sqrt{x^2} = |x| \to |x|^2 = x^2 \ \forall \ x \in R$ 　　(ii) $||x|| = |x| = |-x| \ \forall \ x \in C$

(iii) $|xy| = |x| \cdot |y| \ \forall \ x, y \in C.$

　　The property extends to $|xyzw| = |x||y||z||w|$ & $|x^n| = |x|^n$

(iv) If $|x| = 0$ then $x = 0$ 　　(v) $\left|\dfrac{x}{y}\right| = \dfrac{|x|}{|y|}$, where $x, y \in C$ & $|y| \neq 0$

(vi) Triangular Inequality $\big||x| - |y|\big| \leq |x \pm y| \leq |x| + |y| \ \forall \ x, y \in C$

(vii) If $x, y \in R$ and $|x + y| = |x| + |y|$ then both x and y are of same sign, i.e. $xy \geq 0$.

(viii) If $x, y \in R$ and $|x - y| = |x| + |-y|$ then both x and -y are of same sign, i.e. $xy \leq 0$.

(ix) $|x| = x \to x \geq 0$, the property can be extended to function as well. $|f(x)| = f(x) \to f(x) \geq 0$.

(x) $|x| + x = 0$ or $|x| = -x \to x \leq 0$. Similarly, $|f(x)| = -f(x) \to f(x) \leq 0$.

(xi) If $|x| + |y| = 0 \to x = 0$ and $y = 0 \ \forall \ x, y \in C$

(x) $|x| \geq 0 \ \forall \ x \in R$

(xi) $x^2 + y^2 = 0 \to x = 0, y = 0$ only in real numbers not in complex numbers.

(xii) $1 \leq |\sin(x)| + |\cos(x)| \leq \sqrt{2} \ \forall \ x \in R$

QUESTIONS

1. If $|x - 2| = 5$ & $|y| = 4$ then select the correct options

   (A) $x = -1$ or 7 (B) $y = \pm 4$ (C) $x + y \in \{-5, 3, 11\}$

   (D) Sum of all the different possible values of $|x| + |y|$ is 16.

2. If $|xy| = 6$ & $x, y \in I$ then minimum value of $|x|^{|y|} + y + x$, is

   (A) 4   (B) -5   (C) -6   (D) 1

3. The number of real roots of the equation $|x|^2 - 3|x| + 2 = 0$, is

   (A) 1   (B) 2   (C) 3   (D) 4

4. Find the real roots of the equation $|x|^2 - 2|x| - 3 = 0$.

5. Solve the equation $|2x - 3| = 8$.

6. Find the number of solutions of equation $|xy| = |y|$, hence plot the solution curve on Cartesian Co-ordinate system.

7. Solve the following Equations

   (i) $|x + 2| = 2(3 - x)$ (ii) $|3x - 2| + x = 11$

   (iii) $||x - 1| + 2| = 1$ (iv) $|x - 2|^2 + |x - 2| - 2 = 0$

   (v) $x^2 - |x| - 2 = 0$

8. The number of real roots of the equation $x^2 - 4x - 3|x - 2| + 6 = 0$ is/are?

   (A) 0   (B) 2   (C) 3   (D) 4

9. $|x + 1| + |x - 1| + |x + 2| + |x - 2| = 0$ then x has

   (A) The values $\{1, -1, 2, -1\}$ (B) the values $\left\{\dfrac{1}{2}, -\dfrac{1}{2}, \dfrac{3}{2}, -\dfrac{3}{3}\right\}$

   (C) No solution   (D) none of these

10. If $\dfrac{1}{|x|}+\dfrac{1}{x}=1$ then the value of x is/are

    (A) 1 (B) 2 (C) 3 (D) None of these

11. If ordered pairs (x, y) satisfies $|3x - 1| = 2$ and $|y - 1| = 1$ then the correct option(s) is/are

    (A) Only 3 pairs (x, y) are possible

    (B) 4 pairs (x, y) are possible

    (C) The maximum area formed by line joining pairs = 8/3 sq. unit

    (D) No of possible matrices $\begin{bmatrix} -x & y \\ -y & x \end{bmatrix}$ are 4.

12. The minimum value of $|x-1| + |x-3| + |x-4|$ is

    (A) 5 (B) 4 (C) 3 (D) None of these

13. The number of integral values of x satisfying $|(x-1)(x-7)| > x^2 - 8x + 7$, are

    (A) 5 (B) 4 (C) 3 (D) None of these

14. Solve the following equations

    (i) $|x-1| + |x-2| = 2$ (ii) $|x-1| + |x-2| = 3$

    (iii) $|x| - |x-1| = 1$ (iv) $|2x-1| + |x| = 5$

15. Solve the equation $|2x-1| = |x+2|$

16. Find the maximum and minimum value of $|x+y|$ if $x \in [-1, 2]$ & $y \in [-1, 1]$

17. Find the minimum value of the expression $F(x) = |x-1| + |x-2| + |x-3| + ... + |x-50|$

    Where $x \in R$

18. (i) Solve :

    (a) $|x-1| < 2$ (b) $|x-3| \geq 3$ (c) $|x-5| < -6$ (d) $|x-3| > -5$

    (e) $2 < |x-1| \leq 3$ (f) $2 < |3x-2| < 5$

(ii) $|x-1| \leq 4 \leq |x+1|$, then

(A) $x \in [3, 5]$ (B) $x \in [-3, 5]$ (C) $x \in [-1, 1]$ (D) $x \in (-\infty, -3) \cup (5, \infty)$

19. $3 < |x-1| < 4$ and $-1 < |x-2| < 6$, then $x \in$

(A) $x = \{1, 2, 3\}$ (B) $x = \left\{\dfrac{3}{2}, \dfrac{5}{2}\right\}$ (C) Φ (D) none of these

20. $||x-1|-2| \leq 5$, then $x \in$

(A) $x \in [-3, 7]$ (B) $x \in [-8, 8]$ (C) $x \in [-6, 8]$ (D) none of these

21. $||x-4|-3| \geq 6$, then $x \in$

(A) $x \in (-\infty, -5] \cup [13, \infty)$ (B) $x \in (-\infty, -3] \cup [9, \infty)$

(C) x has no solution (D) only x = {-5} or {13} or {20}

22. Solve the following inequalities

(i) $|x^2 - 2x + 2| \geq 4$ (ii) $|2 - \sqrt{2-x}| \leq 0$

(iii) $3 < |x^2 - x| \leq 4$ (iv) $|x|^3 - 2x^2 - 4|x| + 3 < 0$

(v) $|x^2 + 6x + 7| = |x^2 + 4x + 4| + |2x + 3|$ (vi) $|x^2 - 7x + 12| > x^2 - 7x + 12$

(vii) $|3x - 5| - |2x + 3| > 0$ (viii) $|x^2 - 2x - 3| < 3x - 3$

(ix) $|x^2 - 5x| > |x|^2 - 5|x|$

THANKS!
☺
Keep smiling!

Visit Us: https://www.mathsarc.com

ANSWER KEY

1. B 2. C 3. D 4. x = ± 3

5. x = 11/2 or – 5/2.

7. (i) x = 4/3 (ii) $x = -\dfrac{9}{2}$ and $\dfrac{13}{4}$. (iii) no solution (iv) x = 3 and 1

    (v) x = ± 2

8. D 15. x = 3 and – 1/ 3

Mathsarc Education

A learning place to fulfill your dream of success!

MATHEMATICS   IIT JEE Main/Advanced

LOGARITHM

LOGARITHMIC PLOTS ARE A DEVICE OF THE DEVIL!

SINGLE OPTION CORRECT

1. If $x, y, z \in R+$ are such that $z > y > x > 1$, $\log_y x + \log_x y = 5/2$ and $\log_z y + \log_y z = 10/3$, then $\log_x z$ is equal to –

    (A) 2 (B) 3 (C) 6 (D) 12

2. If x, y, z are three positive numbers in G.P., then $\dfrac{1+\ln x}{2}, \dfrac{1+\ln y}{4}, \dfrac{1+\ln z}{8}$ are in

    (A) A.P. (B) G.P. (C) H.P. (D) A.G.P.

3. The solution set of the inequality $\sqrt{\log_2 x - 1} + \dfrac{1}{2}\log_{1/2}(x^3) + 2 > 0$ is.

    (A) [2, 3) (B) (2, 3] (C) [2, 4) (D) (2, 4]

4. If a, b, c are real numbers greater than 1 and
$x = \dfrac{1}{1+\log_{a^3b^2}(c^2/a)} + \dfrac{1}{1+\log_{b^3c^2}(a^2/b)} + \dfrac{1}{1+\log_{c^3a^2}(b^2/c)}$, then 2x is equal to

5. The solution set of the inequation $\log_{1/3}(x^2 + x + 1) + 1 > 0$ is:

    (A) $(-\infty, -2) \cup (1, \infty)$ (B) [-1, 2] (C) (-2, 1) (D) R

6. Number of integers which satisfy the inequation $\log_2 \sqrt{\dfrac{|x|}{2}} < \log_{\left(\frac{|x|}{4}\right)} 2$ is

    (A) 4 (B) 6 (C) 8 (D) 10

7. If α, β are the solutions of equation $2^{20+\log_2^2 x} = x^{12}$ such that α < β, then $\dfrac{\beta}{\alpha}$ is –

   (A) 4   (B) 64   (C) 256   (D) 512

8. The sum of all the natural numbers for which $\log_{(4-x)}(x^2 - 14x + 45)$ is defined is –

   (A) 1   (B) 2   (C) 3   (D) 4

9. The number of solution(s) of the equation $\log_7(2^x - 1) + \log_7(2^x - 7) = 1$, is

   (A) 0   (B) 1   (C) 2   (D) 3

10. If $55^{f(x)} + 5^x - 2012 = 0$ and f(x) is defined. Then possible integral value(s) of x can't be

    (A) –1   (B) 2   (C) 3   (D) 5

11. Let A denotes the value of $\log_{10}\left(\dfrac{ab+\sqrt{(ab)^2-4(a+b)}}{2}\right) + \log_{10}\left(\dfrac{ab-\sqrt{(ab)^2-4(a+b)}}{2}\right)$

    where a = 43, b = 57 and B denotes the value of the expression $\left(2^{\log_6 18}\right) \times \left(3^{\log_6 3}\right)$.

    Find the value of (A×B).

    (A) 12   (B) 18   (C) 43   (D) None of these

12. The increasing geometric sequence $x_0, x_1, x_2, \ldots\ldots$ consists entirely of integral power of 3. Given that $\sum_{n=0}^{7} \log_3(x_n) = 308$ and $56 \leq \log_3\left(\sum_{n=0}^{7} x_n\right) \leq 57$, then the value of $\log_3(x_{14})$ is equal to ___

    (A) 70   (B) 91   (C) 112   (D) 133

13. Let $f(x) = (x^2 + 3x + 2)^{\cos(\pi x)}$.

    Find the sum of all positive integers n for which $\left|\sum_{k=1}^{n} \log_{10} f(k)\right| = 1$

    (A) 18   (B) 19   (C) 20   (D) 21

14. The system of equations

$$\log_{10}(2000xy) - (\log_{10} x)(\log_{10} y) = 4$$
$$\log_{10}(2yz) - (\log_{10} y)(\log_{10} z) = 1$$
$$\log_{10}(zx) - (\log_{10} z)(\log_{10} x) = 0$$

Has two solutions (x_1, y_1, z_1) and (x_2, y_2, z_2) then the value of $y_1 + y_2$ is ___

(A) 20   (B) 25   (C) 100   (D) 101

MULTIPLE OPTIONS CORRECT

1. If $6^{x+1} = 2^{3x+1}$, then x is equal to –

   (A) $\dfrac{\log_2 3}{2 - \log_2 3}$ (B) $\log_{4/3} 3$ (C) $\log_{3/4} 3$ (D) $\dfrac{1}{1 - 2\log_3 2}$

2. Let $S = \log_a bc + \log_b ca + \log_c ab$ where a, b, c are real numbers greater than 1, then 'S' can be equal to –

   (A) 4   (B) 6   (C) 3   (D) 8

SUBJECTIVE PROBLEMS

1. If $\log_4(x + 2y) + \log_4(x - 2y) = 1$, then the minimum value of $|x| - |y|$ is _____

2. Let a, b, c, d be positive integers and $\log_a b = 3/2$, $\log_c d = 5/4$. If $a - c = 9$, then $b - d =$ _____

3. Evaluate the following

   (i) $\log_{64} 512$ (ii) $\log_{10}(0.01)$ (iii) $\log_{19} 6859$ (iv) $\log_2\left(16^{1/3} \times \sqrt{8}\right)$

   (v) $\log_2\left(\dfrac{3\sqrt{4}}{4^2 \sqrt{8}}\right)$ (vi) $2^{-\dfrac{\log_2 64}{3}}$

4. Simplify the following:

   (i) $4^{5\log_{4\sqrt{2}}(3-\sqrt{6})-6\log_8(\sqrt{3}-\sqrt{2})}$

   (ii) $\dfrac{2}{\log_4(2000)^6}+\dfrac{3}{\log_5(2000)^6}$

   (iii) $5^{\log_{1/5}\left(\frac{1}{2}\right)}+\log_{\sqrt{2}}\dfrac{4}{\sqrt{7}+\sqrt{3}}+\log_{1/2}\dfrac{1}{10+2\sqrt{21}}$

   (iv) $49^{(1-\log_7 2)}+5^{-\log_5 4}$

   (v) $\log_{1/3}\sqrt[4]{729\cdot\sqrt[3]{9^{-1}\times 27^{-4/3}}}$

   (vi) $a^{\frac{\log_b(\log_b N)}{\log_b a}}$

   (vii) $\dfrac{81^{\frac{1}{\log_5 9}}+3^{\frac{3}{\log_{\sqrt{6}} 3}}}{409}\times\left(\left(\sqrt{7}\right)^{\frac{2}{\log_{25} 7}}-(125)^{\log_{25} 6}\right)$

5. Prove that

   (i) $\log(1+2+3)=\log 1+\log 2+\log 3$

   (ii) $\log 360 = 3\log 2 + 2\log 3 + \log 5$

   (iii) $\log\left(\dfrac{50}{147}\right)=\log 2+2\log 5-\log 3-2\log 7$

   (iv) $\log 40 = 3\log 2 + \log 5$

   (v) $\log 2178 = 5\log 3 + \log 9$

   (vi) $\log\dfrac{a^2}{bc}+\log\dfrac{b^2}{ac}+\log\dfrac{c^2}{ab}=0$

   (vii) $7\log\dfrac{16}{25}+5\log\dfrac{25}{24}+3\log\dfrac{81}{80}=\log 2$

   (viii) $\dfrac{\log_2 24}{\log_{96} 2}-\dfrac{\log_2 192}{\log_{12} 2}=3$

   (ix) $3\log\dfrac{49}{16}+\dfrac{5}{2}\log\dfrac{256}{81}-2\log\dfrac{343}{243}=8\log 2$

6. (i) If $x=\log_3 4$ & $y=\log_5 3$. Find the value of $\log_3 10$ & $\log_3(6/5)$ in terms of x and y.

   (ii) If $k^{\log_2 5}=16$, find the value of $k^{(\log_2 5)^2}$.

7. (i) If $\log\dfrac{a+b}{3}=\dfrac{1}{2}(\log a+\log b)$. Show that $a^2+b^2=7ab$.

   (ii) If $x^2+y^2=23xy$ show that $\log\left(\dfrac{x+y}{5}\right)=\dfrac{1}{2}(\log x+\log y)$.

8. Given $\log_{10} 2 = 0.3010$, $\log_{10} 3 = 0.4771$, Evaluate $\log(36)^{1/4}$.

LOGARITHMIC EQUATIONS

1. $\log_{x-1} 3 = 2$

2. $\log_3(3^x - 8) = 2 - x$

3. $\log_3(x+1) + \log_3(x+3) = 1$

4. $\log_7(2^x - 1) + \log_7(2^x - 7) = 1$

5. $\log_3(1 + \log_3(2^x - 7)) = 1$

6. $9^{\log_3(1-2x)} = 5x^2 - 5$

7. $x^{2\log x} = 10x^2$

8. $3\sqrt{\log_2 x} - \log_2 8x + 1 = 0$

9. $\log^2 x - 3\log x = \log(x^2) - 4$

10. $2\log_3 \dfrac{x-3}{x-7} + 1 = \log_3 \dfrac{x-3}{x-1}$

11. $\log_{\frac{1}{3}}(5x - 1) > 0$

12. $\log_3(3x - 1) < 1$

13. $\log_{0.5}(x^2 - 5x + 6) > -1$

14. $\log_8(x^2 - 4x + 3) \leq 1$

15. $\log_7 \dfrac{2x-6}{2x-1} > 0$

ANSWER KEY - Logarithm

SINGLE OPTION CORRECT

1. C 2. D 3. C

10 D

$55^{f(x)} = 2012 - 5^x \Rightarrow f(x) = \log_{55}(2012 - 5^x)$

for f(x) to be defined $2012 - 5^x > 0$

$\Rightarrow 5^x < 2012 \Rightarrow x < 5$ [since $5^4 < 2012 < 5^5$]

$\therefore$ Integral value of x = –1, 2, 3

11. A 12. B 13. D 14. B

MULTIPLE OPTION CORRECT

1. A, B 2. B, D

SUBJECTIVE PROBLEMS

1. $\sqrt{3}$ 2. 93

3. (i) 1.5 (ii) –2 (iii) 3 (iv) 17/6

(v) –29/6 (vi) 1/4

4. (i) 9 (ii) 1/6 (iii) 6 (iv) 25/2

(v) –1 (vi) $\log_b N$ (vii) 1

6. (i) $\dfrac{xy+2}{2y}, \dfrac{xy+2y-2}{2y}$ (ii) 625

8. 0.3891

LOGARITHMIC EQUATIONS

1. $\{1+\sqrt{3}\}$ 2. $\{2\}$ 3. $\{0\}$ 4. $\{3\}$

5. $\{4\}$ 6. $\{-2-\sqrt{10}\}$ 7. $\{\sqrt{10^{1-\sqrt{3}}}, \sqrt{10^{1+\sqrt{3}}}\}$

8. $\{2, 16\}$ 9. $\{10, 10^4\}$ 10. $\{-5\}$ 11. $\left(\dfrac{1}{5}, \dfrac{2}{5}\right)$

12. (1/3, 2) 13. $(1,2) \cup (3,4)$ 14. $[-1,1) \cup (3, 5]$ 15. $\left(-\infty, \dfrac{1}{2}\right)$

Mathsarc Education

A learning place to fulfill your dream of success!

MATHEMATICS IIT JEE Main/Advanced

SEQUENCES AND SERIES

Fibonacci sequence: 1, 1, 2, 3, 5, 8, 13, 21,.. ∞!

SINGLE OPTION CORRECT

1. If the sum of first n terms of an A.P. is cn(n - 1), where $c \neq 0$, then sum of the squares of these terms is

   (A) $c^2 n^2 (n+1)^2$
   (B) $\frac{2}{3} c^2 n(n-1)(2n-1)$
   (C) $\frac{2}{3} c^2 n(n+1)(2n+1)$
   (D) $\frac{c^2 n^2}{3}(n+1)^2$

2. If $S_n = 1 + \frac{1}{2} + \frac{1}{2^2} + \ldots + \frac{1}{2^{n-1}}$, then the least integral value of n such that $2 - S_n < \frac{1}{100}$ is

   (A) 7 (B) 9 (C) 8 (D) 6

3. If $T_n = (n^2 + 1) \cdot n!$ and $S_n = T_1 + T_2 + T_3 + \ldots + T_n$. Let $\frac{T_{10}}{S_{10}} = \frac{a}{b}$ are relatively prime natural numbers, then the value of (b - a) is

   (A) 8 (B) 9 (C) 10 (D) 11

4. If $\sum_{r=1}^{n} r^3 - \sum_{p=1}^{n}\sum_{m=1}^{p}\sum_{r=1}^{m} 1 = 80$, then possible value of n can be –

   (A) 3 (B) 4 (C) 5 (D) 6

5. The value of $\sum_{k=1}^{\infty} \frac{3k^2 + 3k + 1}{(k^2 + k)^3}$ is equal to

   (A) 1/8 (B) 1/4 (C) 1/2 (D) 1

6. If n arithmetic means $A_1, A_2, ..., A_n$ are inserted between 50 and 100 and n harmonic means $H_1, H_2, H_3,..., H_n$ are inserted between the same two numbers, then $A_2 H_{n-1}$ is equal to

   (A) 5000 (B) 10000/n (C) 10000 (D) 250/n

7. Let $a_0 = \dfrac{5}{2}$ and $a_k = a_{k-1}^2 - 2$ for $k \geq 1$, then the value of $\prod_{k=0}^{\infty}\left(1 - \dfrac{1}{a_k}\right)$ is-

   (A) 1/5 (B) 2/5 (C) 3/7 (D) 4/7

8. The value of $\sum_{n=2}^{\infty} \dfrac{n}{1 + n^2(n^2 - 2)}$ is equal to

   (A) 5/4 (B) 1 (C) 5/16 (D) ¼

9. If p times the pth term of an A.P. is equal to q times the qth term of an A.P., then (p + q)th term is?

   (A) 0 (B) 1 (C) 2 (D) 3

10. The geometric series $a + ar + ar^2 + ar^3 + ...\infty$ has sum 7 and the terms involving odd powers of r has sum '3', then the value of $(a^2 - r^2)$ is –

    (A) 5/4 (B) 5/2 (C) 25/4 (D) 5

11. Consider a sequence whose sum of first n-terms is given by $S_n = 4n^2 + 6n$, $n \in N$, then T_{15} of this sequence is –

    (A) 118 (B) 120 (C) 122 (D) 86

12. Let a_n be a sequence such that $a_1 = 5$ and $a_{n+1} = a_n + (n - 2)$ for all $n \in N$, then a_{51} is
    (A) 1165 (B) 1170 (C) 1175 (D) 1180

13. If $b > 0$, then minimum value of $\dfrac{1 + b^2 + b^3 + b^4 + 8b^5 + b^6 + b^7 + b^{13}}{b^5}$ is equal to-

    (A) $8^{9/8}$ (B) 7 (C) 64 (D) 15

14. If G_1 & G_2 are two geometric means & A is A.M inserted between two numbers, then value of $\dfrac{G_1^2}{G_2} + \dfrac{G_2^2}{G_1}$ is

(A) $\dfrac{A}{2}$ (B) A   (C) 2A   (D) 3A

15. If $S = 1 + \dfrac{1}{4} + \dfrac{1}{9} + \dfrac{1}{16} + \ldots \infty$, then $s = 1 + \dfrac{1}{9} + \dfrac{1}{25} + \dfrac{1}{49} + \ldots \infty$ is equal to:

(A) $\dfrac{S}{2}$ (B) $\dfrac{3S}{4}$ (C) $S - \dfrac{1}{4}$ (D) $S - \dfrac{1}{2}$

16. Let $a_1, a_2, a_3, a_4, \ldots, a_{11}$ be a geometric sequence. If $\prod\limits_{k=1}^{11} a_k = 6$, then the value of $(a_5 a_6 a_7)$ is equal to:

(A) $6^{5/11}$ (B) $25^{1/11}$ (C) $216^{1/11}$ (D) $343^{1/11}$

17. The sum of the series, $1 + 2 \cdot \left(1 + \dfrac{1}{n}\right) + 3 \cdot \left(1 + \dfrac{1}{n}\right)^2 + 4 \cdot \left(1 + \dfrac{1}{n}\right)^3 + \ldots \infty$ where $\left|1 + \dfrac{1}{n}\right| < 1$, is:

(A) n^2 (B) $n(n+1)$ (C) $n\left(1 + \dfrac{1}{n}\right)^2$ (D) $(n+1)(n+2)$

18. Let $F_0 = 0$, $F_1 = 1$ and $F_n = F_{n-1} + F_{n-2}$ $(n \geq 2)$, then the value of the sum $\sum\limits_{n=1}^{\infty} \dfrac{F_n}{3^n}$ is:

(A) 3/5   (B) 1/3   (C) 2/3   (D) 7/9

19. The co-efficient of x^{15} in the product $(1-x)(1-2x)(1-2^2 x)(1-2^3 x) \ldots (1-2^{15} x)$ is equal to

(A) $2^{105} - 2^{121}$ (B) $2^{121} - 2^{105}$ (C) $2^{120} - 2^{104}$ (D) $2^{104} - 2^{120}$

20. Let a_n be the nth term of an A.P. If $\sum\limits_{r=1}^{100} a_{2r} = \alpha$ and $\sum\limits_{r=1}^{100} a_{2r-1} = \beta$, then the common difference of the A.P. is

(A) $\alpha - \beta$ (B) $\dfrac{\alpha - \beta}{100}$ (C) $\beta - \alpha$ (D) $\dfrac{\alpha - \beta}{200}$

21. Suppose a, b, c are in A.P. and a^2, b^2, c^2 are in G.P. If $a < b < c$ and $a + b + c = 3/2$, then value of a is

(A) $\dfrac{1}{2\sqrt{2}}$ (B) $\dfrac{1}{2\sqrt{3}}$ (C) $\dfrac{1}{2} - \dfrac{1}{\sqrt{3}}$ (D) $\dfrac{1}{2} - \dfrac{1}{\sqrt{2}}$

22. Let $S_k = \sum_{i=0}^{\infty} \dfrac{1}{(k+1)^i}$, then $\sum_{k=1}^{n} k \cdot S_k$ equal:

   (A) $\dfrac{n(n+1)}{2}$ (B) $\dfrac{n(n-1)}{2}$ (C) $\dfrac{n(n+2)}{2}$ (D) $\dfrac{n(n+3)}{2}$

23. If $x, y, z > 0$ and $x + y + z = 1$ then $\dfrac{xyz}{(1-x)(1-y)(1-z)}$ is necessarily

   (A) ≥ 8 (B) $\leq 1/8$ (C) 1 (D) None of these

24. The sum of the series $1+ 3x+ 6x^2+10x^3+\ldots\infty$ will be (where $|x|<1$)

   (A) $\dfrac{1}{(1-x)^2}$ (B) $\dfrac{1}{1-x}$ (C) $\dfrac{1}{(1+x)^2}$ (D) $\dfrac{1}{(1-x)^3}$

25. The number of common terms to the two sequences 17, 21, 25,…., 417 & 16, 21, 26,…, 466 is

   (A) 21 (B) 19 (C) 20 (D) 91

26. It's given that $\dfrac{1}{1^4}+\dfrac{1}{2^4}+\dfrac{1}{3^4}+\dfrac{1}{4^4}+\ldots\infty = \dfrac{\pi^4}{90}$. Then $\dfrac{1}{1^4}+\dfrac{1}{3^4}+\dfrac{1}{5^4}+\dfrac{1}{7^4}+\ldots\infty$ is equal to

   (A) $\dfrac{\pi^4}{96}$ (B) $\dfrac{\pi^4}{45}$ (C) $\dfrac{89\pi^4}{90}$ (D) none of these

27. If $S_n = \sum_{r=1}^{n}\left(\dfrac{2r+1}{r^4 + 2r^3 + r^2}\right)$ then S_{20} is equal to

   (A) $\dfrac{220}{221}$ (B) $\dfrac{420}{441}$ (C) $\dfrac{439}{221}$ (D) $\dfrac{440}{441}$

28. If $1, \dfrac{x}{2}, y$ are in H.P. $(x, y \neq 0)$, then the number of integral ordered pair (x,y) is

   (A) 8 (B) 3 (C) 4 (D) 5

29. If $\sum_{r=1}^{n}\dfrac{r}{1.3.5.7\ldots(2r+1)} = \dfrac{a}{b}\left(1-\dfrac{c}{1.3.5\ldots(2n+1)}\right)$ Where $a,b,c \in N$, $a<b$ & b is a prime number then the value of $(a+b+c)$ is

   (A) 1 (B) 2 (C) 3 (D) 4

30. Let a, b, c are in AP & $|a|, |b|, |c| < 1$ if $x = 1 + a + a^2 + \ldots$ to ∞, $y = 1 + b + b^2 + \ldots \infty$, $z = 1 + c + c^2 + \ldots \infty$, then x, y, z are in

   (A) AP   (B) GP   (C) HP   (D) None

31. If a, b, c, d, e be 5 numbers such that a, b, c are in AP, b, c, d are in GP, c, d, e are in HP. Then a, c, e are in

   (A) AP   (B) GP   (C) HP   (D) None

32. The first term of an infinitely decreasing GP is unity and its sum is S. The sum of the squares of the terms of the progression is

   (A) $\dfrac{S}{2S-1}$ (B) $\dfrac{S^2}{2S-1}$ (C) $\dfrac{S}{2-S}$ (D) S^2

33. Consider an AP $a_1, a_2, a_3 \ldots$ such that $a_3 + a_5 + a_8 = 11$ and $a_4 + a_2 = -2$, then the value of $a_1 + a_6 + a_7$ is

   (A) -8   (B) 5   (C) 7   (D) 9

34. The harmonic mean of two numbers is 4. Their arithmetic mean A and the geometric mean G satisfy the relation $2A + G^2 = 27$, then the numbers are

   (A) 3 & 6   (B) 2 and 9   (C) 9 and 4   (D) 6 and 4

35. Let $b_1, b_2, \ldots, b_n$ be a geometric sequence such that $b_1 + b_2 = 1$ and $\sum_{k=1}^{\infty} b_k = 2$. Given that $b_2 < 0$, then the value of b_1 is

   (A) $2 - \sqrt{2}$ (B) $1 + \sqrt{2}$ (C) $2 + \sqrt{2}$ (D) $4 + \sqrt{2}$

36. Let 1, 4, 7, ... and 9, 16, 23, ... be two arithmetic progressions. The set S is the union of the first 2019 terms of each sequence. How many distinct numbers are in S?

   (A) 3650   (B) 3750   (C) 3800   (D) 3850

37. Given the progression $10^{1/11}, 10^{2/11}, 10^{3/11}, 10^{4/11}, \ldots, 10^{n/11}$. The least positive integer n such that the product of the first n terms of the progression exceeds 100,000 is

   (A) 8   (B) 9   (C) 10   (D) 11

38. Suppose that $\{a_n\}$ is an arithmetic sequence with $a_1 + a_2 + a_3 + \ldots + a_{100} = 100$ and $a_{101} + a_{102} + \ldots + a_{200} = 200$. What is the value of $a_2 - a_1$?

   (A) 0.0001 (B) 0.001 (C) 0.01 (D) 1

39. The first four terms in an arithmetic sequence are $x + y$, $x - y$, xy and $\dfrac{x}{y}$ in the order. What is fifth term?

   (A) $-\dfrac{15}{8}$ (B) $-\dfrac{6}{5}$ (C) 0 (D) $\dfrac{123}{40}$

40. 150 worker were engaged to finish a piece of work in a certain number of days. Four workers stopped working on the second day, four more workers stopped working on the third day and so on. It took 8 more days to finish the work, then the number of days in which the work was completed is

   (A) 29 days (B) 24 days (C) 25 days (D) none of these

MULTIPLE OPTIONS CORRECT

1. Let $x_1, x_2, \ldots, x_{4001}$ are in harmonic progression and $x_1 x_2 + x_2 x_3 + \ldots + x_{4000} x_{4001} = 10$ and $\dfrac{1}{x_2} + \dfrac{1}{x_{4000}} = 50$, then-

   (A) $\sum_{r=1}^{4001} \dfrac{1}{x_r} = 100025$
   (B) $\sum_{r=1}^{4001} \dfrac{1}{x_r} = 100000$
   (C) $\left| \dfrac{1}{x_{4001}} - \dfrac{1}{x_1} \right| = 30$
   (D) $\left| \dfrac{1}{x_{4001}} - \dfrac{1}{x_1} \right| = 40$

2. Let $S_n = \sum_{r=2}^{n} \dfrac{3^{r-1}(2r-3)}{r(r-1)}$, then

   (A) S_9 is divisible by 4
   (B) S_9 is divisible by 21
   (C) $10 S_{10} + 3 = 3^{10}$
   (D) $7(S_7 + 3)$, $10(S_{10} + 3)$, $13(S_{13} + 3)$ are in GP

3. Let $S_n = \sum_{k=1}^{4n}(-1)^{\frac{k(k+1)}{2}}k^2$. Then S_n can take value(s):

(A) 1056   (B) 1088   (C) 1120   (D) 1332

4. If 9th, 13th and 15th terms of an A.P. are the first three terms of a geometric series whose sum of infinite terms is 80, then which of the following hold(s) good?

(A) Sum of the first 16 terms of the geometric progression is 860.

(B) First term of the arithmetic progression is 80.

(C) First term of the geometric progression is 40.

(D) If d is the common difference of arithmetic progression and r is the common ratio of geometric progression then $dr = -\frac{5}{2}$.

5. Let a_1, a_2, a_3, a_4 & a_5 be such that a_1, a_2 & a_3 are in AP, a_2, a_3 & a_4 are in GP & a_3, a_4 & a_5 are in HP, then a_1, a_3 & a_5 are not in

(A) GP   (B) AP   (C) HP   (D) AGP

6. If a, b, c, d are four distinct positive real numbers in harmonic progression. Then

(A) $a + d > b + d$ (B) $ad > bc$

(C) $a + \frac{1}{a} \geq 2$ (D) If $a+b+c = 6$, then $a^2bc^3\big|_{max} = 108$

7. If $T_1, T_2, T_3, \ldots T_{100}$ are in AP, then

(A) $\frac{T_{15} + T_{27}}{2} = T_{21}$ (B) $T_{15} + T_{28} = T_{18} + T_{25}$

(C) $T_1 + T_2 + \ldots + T_{2k} = k(T_3 + T_{2k-2}), k < 50$ (D) 63th term from last = T_{38}

8. Consider $S_\infty = 1 + \frac{1}{x} + \frac{1}{x^2} + \frac{1}{x^3} + \ldots \infty$, then

(A) $S_\infty = \frac{x}{x-1} \; \forall \, |x| < 1$ (B) $S_\infty = $ Not Define, if $x = 1$

(C) $S_\infty = \dfrac{3}{2}$, if $x = 3$

(D) $S_\infty > 0 \; \forall \; x \in (-\infty, -1) \cup (1, \infty)$

9. If $2x, x+8, 3x+1$ are first three term of an A.P. then which of the following statement is/are correct

(A) $x = 5$
(B) $T_7 = 28$
(C) $S_{10} = 235$
(D) common diff. = 3

10. Ratio of sum of n terms of two distinct AP is $\dfrac{7n+1}{3n-1}$ then

(A) Ratio of their 6th term = 39/16
(B) Ratio of their 1st term = 4

(C) Ratio of their 5th term = 2
(D) None of these

11. If three positive unequal numbers a, b, c are in HP then

(A) $a + c > 2b$
(B) $a^2 + c^2 > 2b^2$
(C) $a^2 + c^2 > 2ac$
(D) $a^2 + c^2 < b^2$

12. For the AP given by $a_1, a_2, a_3, \ldots a_n, a_{n+1}, \ldots$ the equations satisfied are

(A) $a_1 + 2a_2 + a_3 = 0$
(B) $a_1 - 2a_2 + a_3 = 0$

(C) $a_1 + 3a_2 - 3a_3 - a_4 = 0$
(D) $a_1 - 4a_2 + 6a_3 - 4a_4 + a_5 = 0$

13. Consider an infinite geometric series first term 'a' and common ratio 'r'. If the sum is 4 and the second term is $\dfrac{3}{4}$, then

(A) $a = 3$
(B) $a = \dfrac{3}{2}$
(C) $r = \dfrac{1}{4}$
(D) $r = \dfrac{1}{2}$

14. If a, b, c are in HP, then

(A) $\dfrac{a}{c} = \dfrac{a-b}{b-c}$
(B) $\dfrac{a}{b+c}, \dfrac{b}{c+a}, \dfrac{c}{a+b}$ are in HP

(C) $\dfrac{a}{b+c}, \dfrac{b}{c+a}, \dfrac{c}{a+b}$ are in AP
(D) $\dfrac{2}{b} = \dfrac{1}{a} + \dfrac{1}{c}$

15. The sum of the first three consecutive terms of an AP 9, and the sum of their squares is 35. Then the sum to n terms of the series is

(A) $n(n+1)$
(B) n^2
(C) $n(4-n)$
(D) $n(6-n)$

16. The consecutive digits of a three digit number are in G.P. If the middle digit be increased by 2, then they form an A.P. If 792 is subtracted from this, then we get the number constituting of same three digits but in reverse order. Then number is divisible by

    (A) 7 (B) 49 (C) 19 (D) 15

17. $\sum_{m=1}^{\infty}\sum_{n=1}^{\infty} \dfrac{m^2 n}{(n\cdot 3^m + m\cdot 3^n)} = \dfrac{p}{q}$ (where p, q are co-prime) then:

    (A) p + q = 41 (B) p < q (C) p is a perfect square (D) q is a perfect square

18. Given four positive numbers in A.P. If 5, 6, 9 and 15 are added respectively to these numbers, we get a G.P. then which of the following holds?

    (A) The common ratio of G.P. is 3/2 (B) Common ratio of G.P. is 2/3

    (C) First term of G.P. is 8 (D) Common difference of the A.P. is 3

INTEGER TYPE

1. If the sum of first n terms of an AP (having positive terms) is given by $S_n = (1+2T_n)(1-T_n)$ where T_n is the nth term of series then $T_2^2 = \dfrac{\sqrt{a} - \sqrt{b}}{4}$ (a, b ∈ N) Find the value of (a + b)

2. Suppose that all terms of an A.P. are natural numbers. If the ratio of the sum of first seven terms to the sum of first eleven terms is 6:11 and the seventh term lies in between 130 and 140, then the common difference of A.P. is:

3. Let $a_1, a_2, a_3, \ldots, a_{11}$ be real numbers satisfying $a_1 = 15$, $27 - 2a_2 > 0$ and $a_k = 2a_{k-1} - a_{k-2}$ for k = 3, 4,……11. If $\dfrac{a_1^2 + a_2^2 + a_3^2 + \ldots + a_{11}^2}{11} = 90$, then the value of $\dfrac{a_1 + a_2 + a_3 + \ldots + a_{11}}{11}$ is equal to ___

4. If a,b,c are in AP then value of $\dfrac{(a-c)^2}{(b^2 - ac)}$ will be (a ≠ b ≠ c)

5. Let X be the set consisting of the first 2018 terms of the arithmetic progression 1, 6, 11,…… and Y be the set consisting of the first 2018 terms of the arithmetic progression 9, 16, 23,….. Then, the number of elements in the set X ∪ Y is ___

6. The sum of first three consecutive terms of a decreasing GP is 19 & their product is 216. If

the sum of infinite number of terms of the GP is K then $\dfrac{k}{3}$ is equal to _____

7. If a, b and c are three distinct positive real numbers, then the minimum integral value of the expression $\dfrac{b+c}{a}+\dfrac{c+a}{b}+\dfrac{a+b}{c}$ is equal to _____

8. If $3+\dfrac{1}{4}(3+d)+\dfrac{1}{4^2}(3+2d)+\ldots\text{to }\infty=8$, then the value of d is equal to _____

9. Let $A_1, A_2, A_3, \ldots, A_{51}$ are fifty one arithmetic means inserted between the numbers a and b. If value of $\dfrac{b+A_{51}}{b-A_{51}}-\dfrac{A_1+a}{A_1-a}$ is M, then remainder when M is divided by 10 is

10. Find the sum of infinity decreasing GP, the sum of whose first three terms is equal to 7 and the product of those terms is 8.

11. Let $a_1, a_2, a_3 \ldots a_{10}$ be in AP & $h_1, h_2, h_3 \ldots h_{10}$ be in HP. If $a_1 = h_1 = 2$, $a_{10} = h_{10} = 3$, then find the value of $a_4 h_7$.

12. Let $a_1, a_2, a_3, \ldots, a_{100}$ be an arithmetic progression with $a_1 = 3$ and $S_p = \sum_{i=1}^{p} a_i, 1 \leq p \leq 100$. For any integer n with $1 \leq n \leq 20$, let m = 5n. If $\dfrac{S_m}{S_n}$ does not depend on n, then a_2 is:

13. Let $<a_n>$ be an arithmetic sequence such that $\sum_{i=1}^{50} a_{2i-1} = 50$, then $\left|\sum_{j=1}^{50}(-1)^{\frac{j^2+j}{2}} a_{2j-1}\right|$ is equal to

SUBJECTIVE PROBLEMS

1. Let $\{a_n\}$ be a sequence such that $a_1 = 4$ and sum of first n terms is S_n and $S_{n+1} - 3 S_n - 2n - 4 = 0$ $\forall\ n \in N$, Find a_n.

2. For an Integer $n \geq 3$, Let S_n denotes the sum of the products of the integers from 1 to n taken three at a time. (For example $S_3 = 1\times2\times3 = 6$, $S_4 = 1\times2\times3 + 1\times2\times4 + 1\times3\times4 + 2\times3\times4 = 50$ and so on.) Then the value of of S_{10} = _____. Hence prove by Induction that $S_n = \dfrac{1}{48}n^2(n+1)^2(n^2-3n+2)$.

3. Find the sum of the following series

    (i) $1 + 2x + 3x^2 + 4x^3 + \ldots\ldots$upto n terms
    (ii) $1 + 3 + 7 + 15 + 31 + \ldots$upto n terms

(iii) $1 + 5 + 11 + 19 + 29 + \ldots$ upto n terms (iv) $1.2 + 2.3x + 3.4x^2 + 4.5x^3 + \ldots \infty$ ($|x| < 1$)

(v) $1.n + 2.(n-1) + 3.(n-2) + \ldots + n.1$ (vi) $5 + 7 + 11 + 17 + 25 + \ldots$ upto n terms

4. If x, y, z are positive real numbers, such that $x + y + z = a$, then prove that $\dfrac{1}{x} + \dfrac{1}{y} + \dfrac{1}{z} \geq \dfrac{9}{a}$.

5. Prove the following inequalities

   (i) $(a + b + c)(ab + bc + ca) > 9abc$, where $a > 0, b > 0, c > 0$.

   (ii) If $a + b + c = 1$ then $\dfrac{a}{b+c} + \dfrac{b}{a+c} + \dfrac{c}{a+b} \geq \dfrac{3}{2}$, where $a > 0, b > 0, c > 0$.

   (iii) $\dfrac{a_1}{a_2} + \dfrac{a_2}{a_3} + \dfrac{a_3}{a_4} + \ldots + \dfrac{a_{n-1}}{a_n} + \dfrac{a_n}{a_1} > n$, where $a_i > 0$ for $i = 1, 2, 3, \ldots, n$

   (iv) if $a_1 a_2 a_3 \ldots a_n = 1$ then $(1+a_1)(1+a_2)(1+a_3)\ldots(1+a_n) \geq 2^n$, where $a_i > 0$ for $i = 1, 2, 3, \ldots, n$

   (v) $2^n > 1 + n\sqrt{2^{n-1}}, n > 1$ & $n \in N$.

   (vi) $\dfrac{ab}{c^3} + \dfrac{bc}{a^3} + \dfrac{ca}{b^3} > \dfrac{1}{a} + \dfrac{1}{b} + \dfrac{1}{c}$, where $a > 0, b > 0, c > 0$.

   (vii) If $a > 0, b > 0, c > 0$ find the Minimum value of $(a + b + c)\left(\dfrac{1}{a} + \dfrac{1}{b} + \dfrac{1}{c}\right)$.

   (viii) $\dfrac{a^8 + b^8 + c^8}{a^3 b^3 c^3} \geq \dfrac{1}{a} + \dfrac{1}{b} + \dfrac{1}{c}$ where $a > 0, b > 0, c > 0$

   (ix) if a, b, c represents sides of a triangle then prove that:
   $ab + bc + ca < a^2 + b^2 + c^2 < 2(ab + bc + ca)$.

   (x) If $0 < a, b, c < 1$ and $a + b + c = 2$ then prove that $\left(\dfrac{a}{1-a}\right) \times \left(\dfrac{b}{1-b}\right) \times \left(\dfrac{c}{1-c}\right) \geq 8$

   (xi) For positive numbers a, b, c such that $\dfrac{1}{a} + \dfrac{1}{b} + \dfrac{1}{c} = 1$. Find minimum value of $(a-1)(b-1)(c-1)$

   Ans: 8

6. Find the greatest value of $x^3 y^4$ if $2x + 3y = 7$ and $x \geq 0, y \geq 0$.

7. If $a + b + c = 18$, find the maximum value of $a^2 b^3 c^4$ where $a > 0, b > 0, c > 0$.

8. If $x_1, x_2, x_3, \ldots, x_n$ are in H.P. then prove that $x_1x_2 + x_2x_3 + x_3x_4 + \ldots + x_{n-1}x_n = (n-1)x_1x_n$

9. Calculate the following

   (i) $\sum_{r=1}^{n}\left(x^r + \dfrac{1}{x^r}\right)^2$

   (ii) $\prod_{r=1}^{\infty} 6^{\left(\dfrac{1}{2^r}\right)}$

   (iii) $\sum_{r=1}^{n}\left(\dfrac{r}{1+r^2+r^4}\right)$

   (iv) $\sum\sum_{0 \le i < j \le n} ij$

10. If $\sum_{k=1}^{n}(k^2 + 3k + 3)(k+1)! = (2007)(2007)! - 4$ then the value of n must be

11. If nine A.M.'s & nine H.M's are inserted between 2 & 3. Then prove that $A_i + \dfrac{6}{H_i} = 5$ where $i = 1, 2, \ldots 9$ & A_i, H_i are ith AM & HM respectively.

12. Find the sum of the following series

   (i) $\sum_{r=1}^{n} \dfrac{r}{(r+1)!}$

   (ii) $\sum_{k=1}^{360}\left(\dfrac{1}{k\sqrt{k+1} + (k+1)\sqrt{k}}\right)$

   (iii) $\sum_{k=1}^{100}(k^2+1) \cdot k!$

   (iv) $\sum_{n=1}^{9999} \dfrac{1}{(\sqrt{n}+\sqrt{n+1})(\sqrt[4]{n}+\sqrt[4]{n+1})}$

   (v) $\sum_{r=0}^{9} \dfrac{r^2}{r^2 + (9-r)^2}$

   (vi) $\sum_{r=0}^{n-1} \sin(rx)\cos((n-r)x)$

   (vii) $\dfrac{3}{1\cdot 2} \times \dfrac{1}{2} + \dfrac{4}{2\cdot 3} \times \left(\dfrac{1}{2}\right)^2 + \dfrac{5}{3\cdot 4} \times \left(\dfrac{1}{2}\right)^3 + \ldots + \text{to n terms}$

   (viii) $\sum_{n=1}^{\infty} n^2 e^{-n}$

   (ix) $\sum_{n=0}^{1947} \dfrac{1}{2^n + \sqrt{2^{1947}}}$

13. The first term of an A.P. is 5, the last is 45, and the sum 400. Find the number of terms and common difference.

14. If $S_1, S_2, S_3, S_4, \ldots, S_p$ are the sums of n terms of 'p' arithmetic series whose first terms are 1, 2, 3, 4,….. and whose common difference are 1, 3, 5, 7,……

   prove that $S_1 + S_2 + S_3 + S_4 + \ldots + S_p = \dfrac{np}{2}(np+1)$.

15. Let $f(n) = 1 \times 3 \times 5 \times 7 \times .... \times (2n-1)$. Find the remainder when $f(1) + f(2) + f(3) + ... + f(2016)$ is divided by 100.

16. If $\sum_{r=1}^{n} I(r) = n(2n^2 + 9n + 13)$, then find the sum $\sum_{r=1}^{n} \sqrt{I(r)}$.

17. Prove that $0.4\overline{23} = \dfrac{419}{990}$ using infinite series.

18. Find the sum $S = (x+y) + (x^2 + xy + y^2) + (x^3 + x^2y + xy^2 + y^3) + ... +$ to n terms.

19. If n A.M's are inserted between 20 & 80 such that first mean : last mean = 1 : 3 find the common difference of the corresponding AP.

MATRIX MATCH

1. Match the following

	COULUMN – I		COULUMN – II
A	If a, b, c, d, e are in HP then $\dfrac{ab + bc + cd + de}{ae}$ is equal to	P	1
B	Let T_r be the rth term of an HP & If $T_2 = 3$ & $T_3 = 2$ then T_6 is equal to	Q	2
C	Let $a_1, a_2, a_3, ..., a_{10}$ be in AP & $h_1, h_2, h_3,.....h_{10}$ be in HP & if $a_1 = h_1 = 1$ & $a_{10} = h_{10} = 2$, then $a_4 h_7$ is equal to	R	3
D	Let $H_1, H_2, H_3, ..., H_{12}$ are twelve harmonic mean between 3 & 6, then $\dfrac{1}{H_1} + \dfrac{1}{H_2} + \dfrac{1}{H_3} + ... + \dfrac{1}{H_{12}}$ is equal to	S	4

2. Match the following

	COULUMN - I		COULUMN - II
A	If $2^1 \cdot 2^{1/2} \cdot 2^{1/4} \cdot 2^{1/8} \cdots \infty > 2^x$ then greatest integral value of x is	P	8
B	If a, b, c ∈ {0, 1, 2, 3} such that a ≠ b ≠ c then maximum value of $|a-b|+|b-c|+|c-a|$ is	Q	3
C	If $a^2 + 4b^2 + c^2 - 2a - 4b - 4c + 6 = 0$ then the value of abc is, (where a, b, c ∈ R)	R	1
D	If $x^2 - 2x - k = 0$ possess integral roots then k may be	S	6

ANSWER KEY – Sequence & Series

SINGLE OPTION CORRECT

1. B	2. C	3. B	4. B
5. D	6. A	7. C	8. C
9. A	10. B	11. C	12. D
13. A	14. C	15. B	16. C
17. A	18. A	19. A	20. B
21. D	22. D	23. B	24. D
25. C	26. A	27. D	28. D
29. D	30. C	31. B	32. B
33. C	34. A	35. C	36. B
37. D	38. C	39. D	40. C

MULTI OPTIONS CORRECT

1. A, C	2. A, B, D	3. A, D	4. B, C, D
5. B, C, D	6. A, B, C	7. A, B, C, D	8. C, D
9. A, B, C, D	10. A, B	11. A, B, C	12. B, D
13. A, C	14. A, B, D	15. B, D	16. A, B, C
17. A, B, C	18. A, C, D		

INTEGER TYPE

1. 6	2. 9	3. 0	4.
5. 3748	6. 9	7. 7	8. 9
9. 2	10. 8	11. 6	12. 9 or 3
13. 2			

SUBJECTIVE

1. $a_n = 5 \cdot 3^{n-1} - 1 \ \forall \ n \in N.$ 2. 18150

9. (i) $\left(\dfrac{x^{2n}-1}{x^2-1}\right)\left(x^2 + \dfrac{1}{x^{2n}}\right) + 2n$ (ii) 6

(iii) $\dfrac{n^2+n}{2(n^2+n+1)}$ (iv) $\left(\dfrac{\sum_{r=1}^{n}r^3 - \sum_{r=1}^{n}r^2}{2}\right)$

10. 2005

12. (i)   (ii) 18/19   (iii) 100 × 101!   (iv) 9

(v) 5   (vi) $\left(\dfrac{n-1}{2}\right)\sin(nx)$ (viii) $\dfrac{e(e+1)}{(e-1)^3}$

13. n = 16, d = 8/3   15.   16. $\sqrt{\dfrac{3}{2}}\left(n^2+3n\right)$

18. $\dfrac{1}{x-y}\left(\dfrac{x^2(1-x^n)}{1-x} - \dfrac{y^2(1-y^n)}{1-y}\right)$

MATRIX MATCH

1. A→S, B→P, C→Q, D→R.   2. A→R, B→S, C→R, D→P, Q

Mathsarc Education

A learning place to fulfill your dream of success!

MATHEMATICS   IIT JEE Main/Advanced

QUADRATIC EQUATION & INEQUATIONS

An equation means nothing to me unless it expresses a though of God!

SINGLE OPTION CORRECT

1. If the expression $x^2 - (5m-2)x + (4m^2 + 10m + 25)$ can be expressed as a perfect square, then m =

   (A) 8/3 or 4   (B) -8/3 or 4   (C) 4/3 or 8   (D) -4/3 or 8

2. The value of λ for which one root of the equation $x^2 + (1 - 2\lambda)x + (\lambda^2 - \lambda - 2) = 0$ is greater than 3 and the other is less than 3 is given by

   (A) $\lambda < 2$ (B) $2 < \lambda < 5$ (C) $\lambda > 5$ (D) $\lambda > 1$

3. The value of m for the roots of $2x^2 - mx - 8 = 0$ differ by (m – 1) is

   (A) 4, -10/3   (B) -6, 10/3   (C) 6, 10/3   (D) 6, -10/3

4. If α and β ($\alpha < \beta$) are the roots of the equation $x^2 + bx + c = 0$, where $c < 0 < b$ then

   (A) $0 < \alpha < \beta$ (B) $\alpha < 0 < \beta$ (C) $\alpha < \beta < 0$ (D) Cant Say

5. If the equation $k(6x^2 + 3) + rx + (2x^2 - 1) = 0$ and $6k(2x^2 + 1) + px + (4x^2 - 2) = 0$ have both roots common, then the value p/r is

   (A) 1/2   (B) 2   (C) 1   (D) 4

6. If $y = 2 + \cfrac{1}{4 + \cfrac{1}{4 + \cfrac{1}{4 + \ldots \infty}}}$

   (A) y = 6   (B) y = 5   (C) y = $\sqrt{6}$ (D) y = $\sqrt{5}$

7. The value of $\sqrt{8+2\sqrt{8+2\sqrt{8+2\sqrt{8+.....\infty}}}}$ is

   (A) 10   (B) 6   (C) 8   (D) 4

8. If α and β are roots of $4x^2 - 16x + \lambda = 0$ such that α ∈ (1, 2), β ∈ (2, 3), the sum of all the integral values of λ is

   (A) 42   (B) 32   (C) 22   (D) 12

9. If $f(x) = (x - a_1)^2 + (x - a_2)^2 + (x - a_3)^2 + \ldots + (x - a_n)^2$. Find x where f(x) is minimum

   (A) $-\infty$

   (B) $\dfrac{a_1 + a_2 + a_3 + \ldots + a_n}{n}$

   (C) $-\dfrac{a_1 + a_2 + a_3 + \ldots + a_n}{n}$

   (D) none of these

10. If the larger root of equation $x^2 + (2 - a^2)x + (1 - a^2) = 0$ is less than both the roots of the equation $x^2 - (a^2 + 4a + 1)x + a^2 + 4a = 0$, then the range of a, is

    (A) $\left(-\sqrt{2}, \sqrt{2}\right)$ (B) $\left(-\dfrac{1}{4}, \sqrt{2}\right)$ (C) $\left(-\sqrt{2}, \dfrac{1}{4}\right)$ (D) none of these

11. If one solution of the equation $x^3 - 2x^2 + ax + 10 = 0$ is the additive inverse of another, then which one of the following inequalities is true?

    (A) $-40 < a < -30$ (B) $-30 < a < -20$ (C) $-20 < a < -10$ (D) $-10 < a < 0$

12. The value of $f(x) = x^2 + (p - q)x + p^2 + pq + q^2$ for real values of p, q and x

    (A) is always negative   (B) is always positive

    (C) is some time zero for non zero value of x   (D) None of these

13. Solution set for the inequation $\dfrac{x^2 - 1}{x} \leq 2 - x$ is

    (A) $x \in \left(-\infty, \dfrac{1-\sqrt{3}}{2}\right] \cup \left(0, \dfrac{1+\sqrt{3}}{2}\right]$

    (B) $x \in \left[\dfrac{1-\sqrt{3}}{2}, 0\right) \cup \left[\dfrac{1+\sqrt{3}}{2}, \infty\right)$

    (C) $x \in \left[\dfrac{1-\sqrt{3}}{2}, \dfrac{1+\sqrt{3}}{2}\right]$

    (D) None of these

14. The number of distinct real roots of equation $(|x|-1)^{|x-1|-3} = 1$

    (A) 3   (B) 4   (C) 5   (D) None of these

15. Consider the figure of real quadratic $y = Q(x) = ax^2 + bx + c$ as shown.

    Select the **wrong** option (Where $D = b^2 - 4ac$, $i = \sqrt{-1}$)

    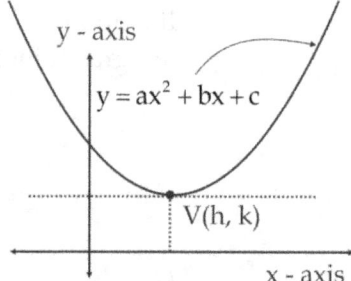

    (A) One root of the equation $ax^2 + bx + c = 0$ is $x = \dfrac{-b + i\sqrt{-D}}{2a}$.

    (B) $ax^2 + bx + c > 0 \; \forall \; x \in R, a \neq 0$

    (C) $|a| + |b| + c = 0$ for at least one real triplet (a, b, c).

    (D) $h = -\dfrac{b}{2a}$ & $k = -\dfrac{D}{4a}$

16. Solution set of $\dfrac{|x-1|}{x(x-2)|x-3|} \geq 0$ is

    (A) $x \in (-\infty, 0) \cup (2, \infty)$

    (B) $x \in (-\infty, 0) \cup (2, \infty) \cup \{1\} - \{3\}$

    (C) $x \in (0, 2)$

    (D) none of these

17. A cubic polynomial P(x) when divided by (x - 1), (x - 2) and (x - 3) leaves remainder 3, 8 and 15 respectively. If P(4) = 30 then the remainder, when P(x) is divided by (x + 1) is

    (A) - 25   (B) - 20   (C) - 16   (D) none of these

18. If $(1+x)(1+x^2)(1+x^4)(1+x^8)......(1+x^{128}) = \sum_{r=0}^{n} x^r$ then n is equal to

    (A) 255   (B) 127   (C) 63   (D) None of these

13. The equation $2^{2x} + (a-1) \cdot 2^{x+1} + a = 0$ has roots of opposite sign then exhaustive set of values of 'a' is

    (A) $a < 0$ (B) $a \in (-1, 0)$ (C) $a \in (-\infty, 1/3)$ (D) $x \in (0, 1/3)$

14. Let α and β are the roots of the equation px² + qx + r = 0, p ≠ 0. If p, q, r are in A.P. and $\frac{1}{\alpha}+\frac{1}{\beta}=4$, then the value of |α - β| is

    (A) $\frac{\sqrt{34}}{9}$
    (B) $\frac{2\sqrt{13}}{9}$
    (C) $\frac{\sqrt{61}}{9}$
    (D) $\frac{2\sqrt{17}}{9}$

15. Solution of the equation: $\sqrt{x+3-4\sqrt{x-1}}+\sqrt{x+8-6\sqrt{x-1}}=1$ is

    (A) x ∈ [4, 9]
    (B) x ∈ [3, 8]
    (C) x ∈ [5, 10]
    (D) x ∈ [4, 7]

16. If the range of $f(x)=\frac{2x^4-14x^2-8x+49}{x^4-7x^2-4x+23}$ is (a, b], then (a + b) is

    (A) 3
    (B) 4
    (C) 5
    (D) 6

17. Consider the equation x² + α x + β = 0 having roots α, β such that α ≠ β. Also consider the inequality $\left||y-\beta|-\alpha\right|<\alpha$, then

    (A) in-equality is satisfied by exactly two integral values of y

    (B) in-equality is satisfied by all values of y ∈ (- 4, 2)

    (C) Roots of the equation are of same sign

    (D) $x^2+\alpha x+\beta>0\,\forall\,x\in[-1,0]$

18. If Q(a) = a² + a+ 1, then number of solutions of equation Q(a²) = 3 Q(a) is

    (A) 0
    (B) 1
    (C) 2
    (D) more than 2

19. If the equation in x, x⁴ + px³ + qx² = 16(2x - 1), where p, q ∈ R has all positive roots, then

    (A) q : |p| = 3
    (B) p > 8
    (C) q > 24
    (D) p < 0 < q < 8

20. Let α, β are the roots of the quadratic equation ax² + bx + c = 0. If a, b, c are in A.P. and α + β = 15, then αβ equals

    (A) - 21
    (B) - 29
    (C) - 31
    (D) - 39

21. Let α and β are the roots of x² - √2 x + 1 = 0, then the value of $\alpha^{50}+\beta^{50}$ is -

    (A) 0
    (B) √2
    (C) 2
    (D) 1

22. If the equation $\dfrac{1}{x}+\dfrac{1}{x-1}+\dfrac{1}{x-2}=3x^3$ has k real roots, then k is equal to -

    (A) 2 (B) 3 (C) 4 (D) 6

23. Let $f(x) = x^3 + x^2 + 1$; $g(x) = x^2 - 1$. If the roots of $f(x)$ are x_1, x_2 and x_3 then the value of $g(x_1)\cdot g(x_2)\cdot g(x_3)+17g(x_1 x_2 x_3)$ is -

    (A) 3 (B) 7 (C) 17 (D) 20

24. Let r(x) be the remainder when the polynomial $x^{135} + x^{125} - x^{115} + x^5 + 1$ is divided by $x^3 - x$. Then

    (A) r(x) is the zero polynomial
    (B) r(x) is a nonzero constant
    (C) degree of r(x) is one
    (D) degree of r(x) is two   (KVPY - 17)

25. Let A, G and H be the arithmetic mean, geometric mean and harmonic mean, respectively of two distinct positive real numbers. If α is the smallest of the two roots of the equation

    $A(G-H)x^2 + G(H-A)x + H(A-G) = 0$, then

    (A) $-2 < α < -1$ (B) $0 < α < 1$ (C) $-1 < α < 0$ (D) $1 < α < 2$

26. The sum of all non-integer roots of the equation $x^5 - 6x^4 + 11x^3 - 5x^2 - 3x + 2 = 0$ is

    (A) 6 (B) -11 (C) -5 (D) 3

27. Let f(x) be a quadratic polynomial with f(2) = 10 and f(-2) = -2. Then the coefficient of x in f(x) is :

    (A) 1 (B) 2 (C) 3 (D) 4

28. Suppose a, b, c are three distinct real numbers. Let $P(x) = \dfrac{(x-b)(x-c)}{(a-b)(a-c)} + \dfrac{(x-c)(x-a)}{(b-c)(b-a)} + \dfrac{(x-a)(x-b)}{(c-a)(c-b)}$ when simplified, P(x) becomes

    (A) 1
    (B) x
    (C) $\dfrac{x^2+(a+b+c)(ab+bc+ca)}{(a-b)(b-c)(c-a)}$
    (D) 0

29. Let a, b, c, d be real numbers such that |a − b| = 2, |b − c| = 3, |c − d| = 4. Then the sum of all possible values of |a - d| is

    (A) 9 (B) 18 (C) 24 (D) 30

30. If $x + \dfrac{1}{x} = a, x^2 + \dfrac{1}{x^3} = b$, then $x^3 + \dfrac{1}{x^2}$ is

    (A) $a^3 + a^2 - 3a - 2 - b$
    (B) $a^3 - a^2 - 3a + 4 - b$
    (C) $a^3 - a^2 + 3a - 6 - b$
    (D) $a^3 + a^2 + 3a - 16 - b$

31. In the real number system, the equation $\sqrt{x+3-4\sqrt{x-1}} + \sqrt{x+8-6\sqrt{x-1}} = 1$ has –

    (A) No solution
    (B) Exactly two distinct solutions
    (C) Exactly four distinct solutions
    (D) Infinitely may solutions

32. Let a, b, c, d be numbers in the set {1, 2, 3, 4, 5, 6} such that the curves $y = 2x^3 + ax + b$ and $y = 2x^3 + cx + d$ have no point in common. The maximum possible value of $(a - c)^2 + b - d$ is

    (A) 0 (B) 5 (C) 30 (D) 36

33. Let f : R → R be the function $f(x) = (x - a_1)(x - a_2) + (x - a_2)(x - a_3) + (x - a_3)(x - a_1)$ with $a_1, a_2, a_3 \in R$. Then $f(x) > 0$ if and only if –

    (A) At least two of a_1, a_2, a_3 are equal
    (B) $a_1 = a_2 = a_3$
    (C) a_1, a_2, a_3 are all distinct
    (D) a_1, a_2, a_3 are all positive and distinct

34. A student notices that the roots of the equation $x^2 + bx + a = 0$ are each 1 less than the roots of the equation $x^2 + ax + b = 0$. Then a + b is:

    (A) - 4 (B) - 2 (C) - 4 (D) - 5

35. Let r be a root of the equation $x^2 + 2x + 6 = 0$. The value of $(r + 2)(r + 3)(r + 4)(r + 5)$ is equal to -

    (A) 51 (B) - 51 (C) - 126 (D) 126

36. Let $p(x) = x^2 - 5x + a$ and $q(x) = x^2 - 3x + b$, where a and b are positive integers. Suppose hcf (p(x), q(x)) = x - 1 and k(x) = lcm (p(x), q(x)). If the coefficient of the highest degree term of k(x) is 1, the sum of the roots of (x - 1) + k(x) is -

    (A) 4 (B) 5 (C) 6 (D) 7

37. Two distinct polynomials f(x) and g(x) are defined as follows: $f(x) = x^2 + ax + 2$; $g(x) = x^2 + 2x + a$. If the equations $f(x) = 0$ and $g(x) = 0$ have a common root then the sum of roots of the equation $f(x) + g(x) = 0$ is -

    (A) -1/2 (B) 0 (C) 1/2 (D) 1

38. Suppose the quadratic polynomial $P(x) = ax^2 + bx + c$ has positive coefficients a, b, c in arithmetic progression in that order. If $P(x) = 0$ has integer roots α and β then $\alpha + \beta + \alpha\beta =$

    (A) 3 (B) 5 (C) 7 (D) 14

39. The number of ordered pairs (x, y) of real numbers that satisfy the simultaneous equations $x + y^2 = x^2 + y = 12$ is

    (A) 0 (B) 1 (C) 2 (D) 4

40. Consider the quadratic equation $a(x-1)^2 + x - 3 = 0$. If a is of the form $\dfrac{k(k+1)}{2}, k \in Q$, then roots of equation are necessarily-

    (A) integers
    (B) imaginary
    (C) rational numbers
    (D) Can not be predicted

41. Set of all real values of 'a' such that $f(x) = \dfrac{(2a-1)x^2 + 2(a+1)x + (2a-1)}{x^2 - 2x + 40}$ is always (-)ve is

    (A) $(-\infty, 0)$ (B) $(0, \infty)$ (C) $(-\infty, 1/2)$ (D) None of these

42. Set of all values of x satisfying the inequality $\sqrt{x^2 - 7x + 6} > x + 2$ is -

    (A) $x \in \left(-\infty, \dfrac{2}{11}\right)$ (B) $x \in \left(\dfrac{2}{11}, \infty\right)$ (C) $x \in (-\infty, 1] \cup [6, \infty)$ (D) $x \in [6, \infty)$

43. Suppose that the roots of $x^3 + 3x^2 + 4x - 11 = 0$ are a, b and c, and the roots of $x^3 + rx^2 + sx + t = 0$ are $a + b$, $b + c$ and $c + a$. Find t __

    (A) 23 (B) 24 (C) 25 (D) 26

44. consider the expression $f(x) = \sin^2 x + 2(1-a)\sin x + a - 1$, $x \in \left(0, \dfrac{5\pi}{6}\right) - \left\{\dfrac{\pi}{2}\right\}$ & a is a real parameter. If $f(x) = 0$ has three solutions then $a \in$

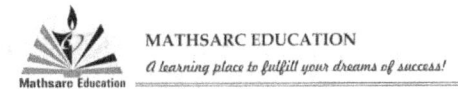

(A) (0, 1)   (B) (1, 3/2)   (C) (3/2, 2)   (D) ∅

MULTIPLE OPTIONS CORRECT

1. The integer value of k for which $(k-2)x^2 + 8x + k + 4 > 0 \,\forall x \in R$ is

   (A) 5   (B) 6   (C) 7   (D) 4

2. Find the value of k for which the graph of the quadratic polynomial

   $P(x) = x^2 + (2x + 3)k + 4(x + 2) + 3k - 5$ intersects the axis of x at two distinct points.

   (A) 1   (B) 2   (C) 5   (D) 4

3. Select the correct statement(s) for solution set of x

   (A) $|2x-1| > -1 \to x \in R$

   (B) $\dfrac{1}{x-1} < x \to x(x-1) > 1 \,\forall\, x > 1$

   (C) $\dfrac{|x|-1}{x(x-2)} < 0 \equiv \dfrac{(x+1)(x-1)}{x(x-2)} < 0$

   (D) $|x-1|(x-2)^2 \leq 0 \to x \in \phi$

4. Select the correct statement(s) for real numbers a, b, c and d.

   (A) If ab = 0 and a = 0 then b ∈ R

   (B) if ab = ac then $\not{a}b = \not{a}c \to b=c \,\forall a \in R$

   (C) $\dfrac{a^2b}{c} \geq 0 \to \dfrac{b}{c} \geq 0$ & a ∈ R

   (D) $\dfrac{a}{b} \geq \dfrac{c}{d} \to ad \geq bc \,\forall\, b, d \in R^+$

5. If $ax^2 + bx + c = 0$ and $cx^2 + bx + a = 0$ (a, b, c ∈ R) have a common non – real roots then

   (A) $-2|a| < b < 2|a|$

   (B) $-2|c| < |b| < 2|c|$

   (C) $a = \pm c$

   (D) $a = c$

6. Consider the equation $x^2 + x - a = 0$, a ∈ N. If equation has real roots then

   (A) a = 2   (B) a = 6   (C) a = 12   (D) a = 20

INTEGER TYPE

1. The number of irrational solutions of the equation $\sqrt{x^2 + \sqrt{x^2+11}} + \sqrt{x^2 - \sqrt{x^2+11}} = 4$ is _____

2. Number of real values of x satisfying the equation $\sqrt{x^2 - 6x + 9} + \sqrt{x^2 - 6x + 6} = 1$ is _____

3. Find the number of integral values of a for which the system of equations
$$\left.\begin{array}{l} x + ay = 3 \\ ax + 4y = 6 \end{array}\right\} \text{ satisfy } x > 1; y > 0.$$

4. Minimum value of $f(x) = |x - 1| + |2x - 1| + |3x - 1| + |4x - 1|$ is p/q where p/q is in lowest form and $p, q \in I^+$ then $p + q$ is _____

5. When the polynomial $5x^3 + Mx + N$ is divided by $x^2 + x + 1$, the remainder is 0.
   Then the value of $|M + N|$ is _____

6. If $a - 2b = 1$ then value of $a^3 - 6ab - 8b^3$ is equal to _____

7. The value of $\sqrt{1 + 5\sqrt{1 + ... + 2013\sqrt{1 + 2014\sqrt{1 + 2015\sqrt{1 + 2016 \times 2018}}}}}$ is _____

8. Let r, s, t are roots of equation $8x^3 + 1001x + 2008 = 0$. Then value of $(r+s)^3 + (s+t)^3 + (t+r)^3$ is 7k3 (where k is at ten's place). Then value of k is _____

9. If both roots of equation $4x^2 - 20px + 25p^2 + 15p - 66 = 0$ are greater than 2, then sum of all possible integral values of p is ____

10. Let k be an integer and p is a prime number such that the quadratic equation $x^2 + kx + p = 0$ has two distinct positive integer solutions. Then the value of $-(p+k)$ is.

11. If the first three consecutive terms of a GP are the real roots of the equation $2x^3 - 19x^2 + 57x - 54 = 0$ and k is the sum of infinite number of the terms of this G.P. Then 2k/9 equals

12. Let $(x + 3)^2 (x + 4)^3 (x + 5)^4 = (x + 1)^9 + a_1 (x + 1)^8 + a_2 (x + 1)^7 + ... + a_9$ then $a_2 - 365$ is ___

SUBJECTIVE PROBLEMS

1. Obtain a polynomial of lowest degree with integral coefficients, whose one of the zero is $\sqrt{5}+\sqrt{2}$.

2. Let $P(x)$ be a polynomial such that $x \cdot P(x-1) = (x-4) \cdot P(x) \,\forall x \in R$. Find all such polynomials

3. Let $P(x)$ be a monic cubic equation such that $P(1) = 1$, $P(2) = 2$, $P(3) = 3$ then find $P(4)$.

4. Show that $f(x) = x^{1000} - x^{500} + x^{100} + x + 1$ has no rational roots.

5. Evaluate: $\sqrt[8]{2207 - \cfrac{1}{2207 - \cfrac{1}{2207 - \ldots\infty}}}$. Express your answer in the form $\dfrac{a+b\sqrt{c}}{d}$, where a, b, c, d are integers.

MATRIX MATCH

1. Match the following

	COULUMN - I		COULUMN - II
A	If a, b, c and d are four non-zero real number such that $(d+a-b)^2 +(d+b-c)^2 = 0$ and the roots of the equation $a(b-c)x^2 + b(c-a)x + c(a-b) = 0$ are real and equal then	P	$a+b+c=0$
B	If the roots of the equation $(a^2+b^2)x^2 - 2b(a+c) + (b^2+c^2) = 0$ are real and equal, then	Q	a,b,c are in A.P.
C	If the equation $ax^2+bx+c=0$ and $x^3-3x^2+3x-1=0$ have a common real root, then	R	a,b,c, are in G.P.
D	Let a,b,c be positive real numbers such that the expression $bx^2 + \left(\sqrt{(a+c)^2+4b^2}\right)x + (a+c)$ is non-negative $\forall x \in R$, then	S	a,b,c are in H.P.

COMPREHENSION for Q 1 - 3

The first four terms of a sequence are given by $T_1 = 0$, $T_2 = 1$, $T_3 = 1$, $T_4 = 2$. The general term is given by $T_n = A\, \alpha^{n-1} + B\, \beta^{n-1}$ where A, B, α, β are independent of n and A is positive.

1. The value of $(\alpha^2 + \beta^2 + \alpha\beta)$ is equal to

    (A) 1 (B) 2 (C) 5 (D) 4

2. The value of $5(A^2 + B^2)$ is equal to

    (A) 2 (B) 4 (C) 6 (D) 8

3. The quadratic equation whose roots are α and β is given by

    (A) $x^2 - 2x - 1 = 0$ (B) $x^2 - 2x - 2 = 0$ (C) $x^2 - x - 1 = 0$ (D) None

ANSWER KEY – Quadratic Equations & Inequations

SINGLE OPTION CORRECT

1. D
2. B
3. D
4. B
5. B
6. D
7. D
8. A
9. B
10.
11.
12.
13. A
14. B
15. C
16. B

17. A, Hint: $P(x) = (x - 1)(x - 2)(x - 3) + x^2 + 2x$

18. A
13. C
14. B
15. C

16. C, Hint: $f(x) = 2 + \dfrac{3}{\left(x^2 - 4\right)^2 + (x-2)^2 + 3}$

17. A
18. C

19. A, Hint: AM of roots = HM → $\alpha = \beta = \gamma = \delta = 2$. P = − 8 and q = 24

20. C
21. A
22. C
23. A
24. C
25. B
26. D, Hint: $(x - 1)(x - 2)(x^3 - 3x + 1) = 0$
27. C
28. A
29. B
30. A
31. D
32. B
33. B
34. C
35. C
36. D
37. C
38. C
39. D
40. C
41. A
42. A
43. A
44. D

MULTI OPTIONS CORRECT

1.
2.
3. A, B, C
4. A, C, D
5. A, B, D
6. A, B, C, D

INTEGER TYPE

1.
2.
3.
4. 7

5. 5
6. 1
7. 6
8. 5

9. 7
10. 1
11. 3

12. 371

Hint: $x + 1 = y \to (y + 2)^2 (y + 3)^3 (y + 4)^4 = y^9 + a_1 y^8 + a_2 y^7 + \ldots + a_9$

a_2 = sum of roots taking two at a time.

SUBJECTIVE

1. $P(x) = a(x^4 - 14x^2 + 9)$, where $a \in I$, $a \neq 0$.

2. $P(x) = c\,x(x - 1)(x - 2)(x - 3)$, $c \neq 0$

3. 10

5. $\dfrac{3+\sqrt{5}}{2}$

MATRIX MATCH

1. A→ R, B→ R, C→ P, D → Q

COMPREHENSION for Q 1 - 3

1. B
2. A
3. C

Mathsarc Education

A learning place to fulfill your dream of success!

MATHEMATICS   IIT JEE Main/Advanced

TRIGONOMETRIC RATIO AND IDENTITIES

Trigonometry enable us to measure distance between two stars

SINGLE OPTION CORRECT

1. The maximum value of $\sin(\cos(\tan x))$ is

   (A) $\dfrac{\sqrt{3}}{2}$ (B) $\sin 1$ (C) 1 (D) $\sin(\cos 1)$

2. The value of $\dfrac{1}{4}\tan\left(\dfrac{\pi}{8}\right) + \dfrac{1}{8}\tan\left(\dfrac{\pi}{16}\right) + \dfrac{1}{16}\tan\left(\dfrac{\pi}{32}\right) + \ldots \infty$ terms is equal to

   (A) $\dfrac{5}{\pi} - \dfrac{1}{2}$ (B) $\dfrac{3}{\pi} + \dfrac{1}{2}$ (C) $\dfrac{2}{\pi} - \dfrac{1}{2}$ (D) $\dfrac{4}{\pi} - \dfrac{1}{4}$

3. The value of expression $\displaystyle\sum_{\theta=0}^{8} \dfrac{1}{1+\tan^3(10\theta°)}$ is

   (A) 5 (B) $21/4$ (C) $14/3$ (D) $9/2$

4. $\tan\dfrac{2\pi}{5} - \tan\dfrac{\pi}{15} - \sqrt{3}\tan\dfrac{2\pi}{5}\tan\dfrac{\pi}{15}$ is equal to -

   (A) $-\sqrt{3}$ (B) $\dfrac{1}{\sqrt{3}}$ (C) 1 (D) $\sqrt{3}$

5. The value of $\cot 70° + 4\cos 70°$ is

   (A) $1/\sqrt{3}$ (B) $\sqrt{3}$ (C) $2\sqrt{3}$ (D) $1/2$

6. if $0 < \theta < \beta < \gamma < \pi/2$, then $\dfrac{\sin\theta + \sin\beta + \sin\gamma}{\cos\theta + \cos\beta + \cos\gamma}$ lies between -

   (A) $\cos\theta$ & $\cos\gamma$ (B) $\sin\theta$ & $\sin\gamma$ (C) $\tan\theta$ & $\tan\gamma$ (D) $\cot\theta$ & $\cot\gamma$

7. Minimum value of f (x) = 256 sin²x + 324 cosec²x, ∀ x ∈ R is

   (A) 576　　　(B) 580　　　(C) 584　　　(D) None of these

8. If $E = 6\sin^2 x + 4\sin 2x - 3$, then

   (A) $|E| \leq 5$　　　(B) $|E| > 5$　　　(C) $|E| \in [6, 8]$　　　(D) $|E| \in [8, 10]$

9. The line joining (5, 0) to (10 cosθ, 10 sinθ) is divided internally in the ratio 2 : 3 at P. If θ varies then the locus of P is

   (A) Straight line　　　(B) Pair of straight lines　　　(C) Circle　　　(D) None of these

10. if α and β are solutions of sin²x + a sinx + b = 0 as well as that of cos²x + c cosx + d = 0, then sin(α + β) is equal to

    (A) $\dfrac{a^2 + c^2}{2ac}$　　　(B) $\dfrac{2ac}{a^2 + c^2}$　　　(C) $\dfrac{b^2 + d^2}{2bd}$　　　(D) $\dfrac{2bd}{b^2 + d^2}$

11. The sum $1 + \binom{n}{1}\cos\theta + \binom{n}{2}\cos 2\theta + \binom{n}{3}\cos 3\theta + \ldots + \binom{n}{n}\cos n\theta$ equals

    (A) $\left(2\cos\dfrac{\theta}{2}\right)^n \cos\dfrac{n\theta}{2}$　　　(B) $\left(2\cos^2\dfrac{\theta}{2}\right)^n$　　　(C) $\left(2\cos^2\dfrac{n\theta}{2}\right)^n$　　　(D) None of these

12. The ratio of the circumference of a circle to the perimeter of the inscribed regular polygon with n sides is

    (A) $2\pi : 2n\sin\dfrac{\pi}{n}$　　　(B) $2\pi : n\sin\dfrac{\pi}{n}$　　　(C) $2\pi : 2n\sin\dfrac{2\pi}{n}$　　　(D) $2\pi : n\sin\dfrac{2\pi}{n}$

13. If the area of the circumcircle of a regular polygon with n sides is A then the area of the circle inscribed in the polygon is

    (A) $A\cos^2\dfrac{2\pi}{n}$　　　(B) $\dfrac{A}{2}\left(\cos\dfrac{2\pi}{n} + 1\right)$　　　(C) $\dfrac{A}{2}\cos^2\dfrac{\pi}{n}$　　　(D) $A\left(\cos\dfrac{2\pi}{n} + 1\right)$

14. The value of $\sin 25° \sin 35° \sin 85°$ is equal to

    (A) $\dfrac{\sqrt{3}}{4}$　　　(B) $\dfrac{1}{4}\sqrt{2 - \sqrt{3}}$　　　(C) $\dfrac{5\sqrt{3}}{9}$　　　(D) $\dfrac{1}{4}\sqrt{\dfrac{1}{2} + \dfrac{\sqrt{3}}{4}}$

15. For all angles A, $\dfrac{\sin 2A \cos A}{(1+\cos 2A)(1+\cos A)}$ equals

(A) $\sin \dfrac{A}{2}$ 　　(B) $\cos \dfrac{A}{2}$ 　　(C) $\tan \dfrac{A}{2}$ 　　(D) $\sin A$

16. The value of $\cos^4 \dfrac{\pi}{8} + \cos^4 \dfrac{3\pi}{8} + \cos^4 \dfrac{5\pi}{8} + \cos^4 \dfrac{7\pi}{8}$ is

(A) $3/4$ 　　(B) $\dfrac{1}{\sqrt{2}}$ 　　(C) $3/2$ 　　(D) $\dfrac{\sqrt{3}}{2}$

17. The expression $\tan\theta + 2\tan(2\theta) + 2^2\tan(2^2\theta) + \ldots + 2^{14}\tan(2^{14}\theta) + 2^{15}\cot(2^{15}\theta)$ is equal to

(A) $2^{16}\tan(2^{16}\theta)$ 　　(B) $\tan(\theta)$

(C) $\cot(\theta)$ 　　(D) $2^{16}[\tan(2^{16}\theta) + \cot(2^{16}\theta)]$

18. The value of $2\sin\left(\dfrac{\theta}{2}\right)\cos\left(\dfrac{3\theta}{2}\right) + 4\sin(\theta)\sin^2\left(\dfrac{\theta}{2}\right)$ equals

(A) $\sin\left(\dfrac{\theta}{2}\right)$ 　　(B) $\sin\left(\dfrac{\theta}{2}\right)\cos(\theta)$ 　　(C) $\sin(\theta)$ 　　(D) $\cos(\theta)$

19. If $\cos x + \cos y + \cos z = 0$, $\sin x + \sin y + \sin z = 0$, then $\cos\dfrac{(x-y)}{2}$ is

(A) $\pm\dfrac{\sqrt{3}}{2}$ 　　(B) $\pm\dfrac{1}{2}$ 　　(C) $\pm\dfrac{1}{\sqrt{2}}$ 　　(D) 0

20. If $2\sec 2\theta = \tan\beta + \cot\beta$, then one possible value of $\theta + \beta$ is

(A) $\dfrac{\pi}{2}$ 　　(B) $\dfrac{\pi}{4}$ 　　(C) $\dfrac{\pi}{3}$ 　　(D) 0

21. The value of $\dfrac{\cos 37° + \sin 37°}{\cos 37° - \sin 37°}$ equals

(A) $\tan 8°$ 　　(B) $\cot 8°$ 　　(C) $\sec 8°$ 　　(D) $\operatorname{Cosec} 8°$

22. $P = \tan(3^{n+1}Q) - \tan(Q)$ and $Q = \sum_{r=0}^{n} \dfrac{\sin(3^r Q)}{\cos(3^{r+1} Q)}$ then

   (A) P = 2Q (B) P = 3Q (C) 2P = Q (D) 3P = Q

23. If $(1+\tan 5°)(1+\tan 10°)(1+\tan 15°)........(1+\tan 45°) = 2^k$, then 'k' equals

   (A) 4 (B) 5 (C) 8 (D) 9

24. Find the smallest natural 'n' such that $\tan(107n)° = \dfrac{\cos 96° + \sin 96°}{\cos 96° - \sin 96°}$

   (A) n = 2 (B) n = 3 (C) n = 4 (D) n = 5

25. The value of the expression $\left(1+\cos\dfrac{\pi}{10}\right)\left(1+\cos\dfrac{3\pi}{10}\right)\left(1+\cos\dfrac{7\pi}{10}\right)\left(1+\cos\dfrac{9\pi}{10}\right)$ is

   (A) 1/8 (B) 1/16 (C) 1/4 (D) 0

26. If $\theta \in \left[\dfrac{\pi}{2},\pi\right]$ then the value of $\sqrt{1+\sin\theta} - \sqrt{1-\sin\theta}$ is equal to

   (A) $2\cos\dfrac{\theta}{2}$ (B) $2\sin\dfrac{\theta}{2}$ (C) 2 (D) None of these

27. If $u_n = \cos^n\theta + \sin^n\theta$, then value of $2u_6 - 3u_4$ is?

   (A) 0 (B) 1 (C) -1 (D) 2

28. If $0 < \beta < \alpha \leq \pi/4$, $\cos(\alpha + \beta) = 3/5$ and $\cos(\alpha - \beta) = 4/5$, then $\sin(2\alpha)$ is equals to ___

   (A) 1 (B) 0 (C) 2 (D) None of these

29. If $\tan\theta \cdot \tan\left(\dfrac{\pi}{3}+\theta\right) \cdot \tan\left(\dfrac{\pi}{3}-\theta\right) = -1$, $(0 < \theta < \pi/2)$, then value of $3\sin\theta - 4\cos^3\theta =$

   (A) 1 (B) -1 (C) $1/\sqrt{2}$ (D) $-1/\sqrt{2}$

30. The value of $\dfrac{4\sin 9° \sin 21° \sin 39° \sin 51° \sin 69° \sin 81°}{\sin 54°}$ is equal to

   (A) 1/16 (B) 1/32 (C) 1/8 (D) 1/4

MULTIPLE OPTIONS CORRECT

1. Let $x_0 > 0$. For every natural number n define

   $s_n = \sin\left(\dfrac{\pi x_0^n}{1+x_0^{2n}}\right)$ and $c_n = \cos\left(\dfrac{\pi x_0^n}{1+x_0^{2n}}\right)$. Then for all n

   (A) $s_n^2 + c_n^2 = 1$ (B) $s_n \geq s_{n+1}$ (C) $c_n \geq c_{n+1}$ (D) $\dfrac{s_n}{c_n} \geq \dfrac{s_{n+1}}{c_{n+1}}$

2. $y = \dfrac{1}{3(1+\sin x) - \cos^2 x}$

   (A) Maximum value of y is not defined   (B) Minimum value of y is 1/6

   (C) Maximum value of y is 4/7 at sin x = 0   (D) y is always positive

3. The value of $\sec^4 x + \csc^4 x + \sec^4 x \csc^4 x$ can be

   (A) 18   (B) 24   (C) 32   (D) 64

4. The expression $\cos^2(\theta + b) + \cos^2(\theta + a + b) + \cos^2(a) - 2\cos(a)\cos(\theta + b)\cos(\theta + a + b)$ is independent of ___

   (A) a   (B) b   (C) θ (D) a and b only

5. If $\alpha = \sin x \cos^3 x$ and $\beta = \cos x \sin^3 x$, then:

   (A) $\alpha - \beta > 0$; for all x in $\left(0, \dfrac{\pi}{4}\right)$ (B) $\alpha - \beta < 0$; for all x in $\left(0, \dfrac{\pi}{4}\right)$

   (C) $\alpha + \beta > 0$; for all x in $\left(0, \dfrac{\pi}{2}\right)$ (D) $\alpha + \beta < 0$; for all x in $\left(0, \dfrac{\pi}{2}\right)$

6. If cosx = tanx, then which of the following is true?

   (A) $\dfrac{1}{\sin x} + \cos^4 x = 1$ (B) $\dfrac{1}{\sin x} + \cos^4 x = 2$

   (C) $\cos^4 x + \cos^2 x = 1$ (D) $\cos^4 x + \cos^2 x = 2$

7. The expression $\dfrac{\sec^4\theta}{\tan^2\beta}+\dfrac{\sec^4\beta}{\tan^2\theta}$ (wherever defined) can take the value-

   (A) 4 (B) 6 (C) 8 (D) 10

8. If $\sin^2(2x) + \cos^2(3y) + \tan^2(4z) + \sin(2x).\cos(3y) + \cos(3y).\tan(4z) + \tan(4z).\sin(2x) \leq 0$, where $x, y, z \in \left(0,\dfrac{\pi}{2}\right]$, then possible value of $(x + y + z)$ is-

   (A) $\dfrac{11\pi}{12}$ (B) $\dfrac{7\pi}{6}$ (C) $\dfrac{5\pi}{4}$ (D) $\dfrac{3\pi}{2}$

9. Which of the following functions have the maximum value unity?

   (A) $\sin^2 x - \cos^2 x$
   (B) $\dfrac{\sin 2x - \cos 2x}{\sqrt{2}}$
   (C) $-\dfrac{\sin 2x - \cos 2x}{\sqrt{2}}$
   (D) $\sqrt{\dfrac{6}{5}}\left(\dfrac{1}{\sqrt{2}}\sin x + \dfrac{1}{\sqrt{3}}\cos x\right)$

10. If $\dfrac{\sin^4 x}{3}+\dfrac{\cos^4 x}{5}=\dfrac{1}{8}$, then which of the following holds good?

    (A) $\tan^2 x = \dfrac{5}{3}$
    (B) $\tan^2 x = \dfrac{3}{5}$
    (C) $\dfrac{\sin^6 x}{9}+\dfrac{\cos^6 x}{25}=\dfrac{1}{64}$
    (D) $\dfrac{\sin^6 x}{9}+\dfrac{\cos^6 x}{25}=\dfrac{1}{8}$

11. If $\tan(\pi\cos\theta) = \cot(\pi\sin\theta)$, then the value of $\cos\left(\theta-\dfrac{\pi}{4}\right)$ is equal to

    (A) $\dfrac{1}{2\sqrt{2}}$ (B) $-\dfrac{1}{2\sqrt{2}}$ (C) $\dfrac{3}{2\sqrt{2}}$ (D) $-\dfrac{1}{\sqrt{2}}$

12. For all $\theta \in (0,\pi/2)$, select the correct options

    (A) $\cos(\sin\theta) > \sin(\cos\theta)$
    (B) $\cos(\sin\theta) < \sin(\cos\theta)$
    (C) $\sin\theta > \sin(\sin\theta)$
    (D) $\cos(\cos\theta) > \sin(\sin\theta)$

13. Select the correct options

   (A) $\sin(\sin\theta)|_{max} = \sin(1) \ \forall\theta\in R$

   (B) $\sin(\sin\theta)|_{min} = -1 \ \forall\ \theta\in R$

   (C) $\sin(\sin\theta)|_{min} = -\sin(1) \ \forall\ \theta\in R$

   (D) $\cos(\cos\theta)|_{min} = \cos(1) \ \forall\ \theta\in R$

INTEGER TYPE

1. Roots of $x^3 + ax^2 + bx + c = 0$ are cosines of angles of an acute triangle, then the value of $a^2 - 2b - 2c$ is

2. If $\lim\limits_{\theta\to 0}\left(\dfrac{1}{\theta^3}\left(\sin^3\dfrac{\theta}{3} + 3\sin^3\dfrac{\theta}{3^2} + 3^2\sin^3\dfrac{\theta}{3^3} + 3^3\sin^3\dfrac{\theta}{3^4} + ...\text{upto}\infty \text{ terms}\right)\right) = \dfrac{1}{k!}$ then k is

3. The value of $\sin^2 12° + \sin^2 21° + \sin^2 39° + \sin^2 48° - \sin^2 9° - \sin^2 18°$ is____

4. The value of $\prod\limits_{k=1}^{n}\left(1 + 2\cos\left(\dfrac{2\pi\cdot 3^k}{3^n + 1}\right)\right) = ...$

5. The value of $\sum\limits_{r=0}^{9}\sin^2\dfrac{\pi r}{10}$ is equal to ____

6. Let x and y be positive real numbers and θ an angle such that $\theta \ne \dfrac{n\pi}{2}$ n for any integer n. Suppose $\dfrac{\sin\theta}{x} = \dfrac{\cos\theta}{y}$ and $\dfrac{\cos^4\theta}{x^4} + \dfrac{\sin^4\theta}{y^4} = \dfrac{97\sin 2\theta}{x^3 y + xy^3}$, then the value of $\dfrac{x}{y} + \dfrac{y}{x}$ is ____

7. If the greatest value of $\sin^2\alpha \cos^6\alpha$ is $\dfrac{a}{256}$, then the value of $\left[\dfrac{a}{10}\right]$ is (where [.] denotes greatest integer function)

8. The value of $\cot 76° \cot 44° + \cot 16° \cot 44° - \cot 76° \cot 16°$ is?

9. If $\dfrac{1-\cos a - \tan^2\left(\dfrac{a}{2}\right)}{\sin^2\left(\dfrac{a}{2}\right)} = \dfrac{k\cos a}{w + p\cos a}$ where k, w and p have no common factor other than 1, then the value of $k^2 + w^2 + p^2$ is equal to

10. If y = (sinx + cosecx)² + (cosx + secx)² + (tanx + cotx)² where x ∈ R −{x: x = nπ, (2n+1)π/2 ∀ n ∈ I} and y = p then $\left[\dfrac{p}{3}\right]$ is where [.] = GIF.

11. If α, β, γ, δ be first 4 positive solutions of sinx = 1/4 with α < β < γ < δ, then value of $\sin\dfrac{\delta}{2} + 2\sin\dfrac{\gamma}{2} + 3\sin\dfrac{\beta}{2} + 4\sin\left(\dfrac{\alpha}{2}\right)$ is ____

MATRIX MATCH

1. Match the following

	COULUMN – I		COULUMN – II
A	Maximum value of $\dfrac{1-\tan^2\left(\dfrac{\pi}{4}-x\right)}{1+\tan^2\left(\dfrac{\pi}{4}-x\right)}$	P	1
B	Minimum value of $\log_3\left(\dfrac{5\sin x - 12\cos x + 26}{13}\right)$	Q	0
C	Minimum value of y = − 2sin²x + cosx + 3	R	7/8
D	Maximum value of y = 4sin²θ + 4sinθ cosθ + cos²θ	S	5
		T	6

2. Match the following

	COULUMN – I		COULUMN – II
A	$\dfrac{\prod_{k=1}^{100}\left(1-\tan^2\dfrac{2^k\pi}{2^{100}+1}\right)}{2^{100}}$ is equal to	P	1
B	$\prod_{k=1}^{100}\left(1+2\cos\dfrac{2\pi\cdot 3^k}{3^{100}+1}\right)$ is equal to	Q	−1
C	Let n be +ve integer and let x be a real number different	R	2

	from $2^{k+1}\left(\dfrac{\pi}{3}+l\pi\right), k=1,2,3,\ldots,n, l$ an integer then $\left[\left(1+2\cos\dfrac{x}{2^{100}}\right)\prod_{k=1}^{100}\left(1-2\cos\dfrac{x}{2^k}\right)\right]-2\cos x$ is equal to		
D	The value of $\displaystyle\sum_{n=1}^{\infty}\left(\dfrac{1}{2^n}\tan\dfrac{2}{2^n}\right)+\cot 2$ is equal to	S	3
		T	1/2

3. Match the following

	COULUMN - I		COULUMN - II
A	If $\tan\theta=\dfrac{1}{2}$, and $\tan\phi=\dfrac{1}{3}$, then the value of $\theta+\phi$ is	P	$\dfrac{5\pi}{6}$
B	The value of θ satisfying the equation, $\cos\theta+\sqrt{3}\sin\theta=0$, in $0\le\theta\le\pi$, is	Q	$\dfrac{\pi}{3}$
C	The values of x, satisfying $2\sin^2 x - 3\sin x + 1 \ge 0$ is/are	R	$\dfrac{2\pi}{3}$
D	If $1+\sin\theta+\sin^2\theta+\sin^3\theta+\ldots$ to $\infty = 4+2\sqrt{3}$, $0<\theta<\pi$, $\theta\ne\pi/2$ then θ is equal to	S	$\dfrac{\pi}{4}$

COMPREHENSION

Paragraph for Question No's 1 to 3

In a $\triangle ABC$, if $\cos A\cdot\cos B\cdot\cos C=\dfrac{\sqrt{3}-1}{8}$ & $\sin A\cdot\sin B\cdot\sin C=\dfrac{3+\sqrt{3}}{8}$, then

1. The value of tan A + tan B + tan C is

   (A) $\dfrac{3+\sqrt{3}}{\sqrt{3}-1}$ (B) $\dfrac{4+\sqrt{3}}{\sqrt{3}-1}$ (C) $\dfrac{6-\sqrt{3}}{\sqrt{3}-1}$ (D) $\dfrac{\sqrt{3}+\sqrt{2}}{\sqrt{3}-1}$

2. The value of tan A tan B + tan B tan C + tan C tan A, is

   (A) $5 - 4\sqrt{3}$ (B) $5 + 4\sqrt{3}$ (C) $5 - \sqrt{3}$ (D) $5 + \sqrt{3}$

3. The angles of $\triangle ABC$ are:

   (A) $45°, 30°, 150°$ (B) $45°, 60°, 75°$ (C) $45°, 45°, 90°$ (D) None of these

Paragraph for Question No's 4 to 5

Given $\cos 2^m \theta \cos 2^{m+1} \theta \ldots \cos 2^n \theta = \dfrac{\sin(2^{n+1}\theta)}{2^{n-m+1}\sin(2^m\theta)}$, where $2^m\theta \neq k\pi, n, m, k \in I$.

4. $\sin\dfrac{9\pi}{14} \cdot \sin\dfrac{11\pi}{14} \cdot \sin\dfrac{13\pi}{14}$ is equal to

   (A) $1/64$ (B) $-1/64$ (C) $1/8$ (D) $-1/8$

5. $\cos\left(\dfrac{2^3\pi}{10}\right) \cdot \cos\left(\dfrac{2^4\pi}{10}\right) \cdot \cos\left(\dfrac{2^5\pi}{10}\right) \cdots \cos\left(\dfrac{2^{10}\pi}{10}\right)$ is equal to

   (A) $1/128$ (B) $1/256$ (C) $\dfrac{1}{512}\sin\left(\dfrac{\pi}{10}\right)$ (D) $\dfrac{\sqrt{501}}{512}\sin\left(\dfrac{3\pi}{10}\right)$

THANKS!

Keep smiling!

Visit Us: https://www.mathsarc.com

ANSWER KEY & SOLUTION

SINGLE OPTION CORRECT

1. B
2. C

Hint: $\cot x = \dfrac{1}{2}\left(\cot \dfrac{x}{2} - \tan \dfrac{x}{2}\right) = \dfrac{1}{2}\left\{\dfrac{1}{2}\left(\cot \dfrac{x}{4} - \tan \dfrac{x}{4}\right) - \tan \dfrac{x}{2}\right\}\ldots$

3. A
4. D
5. B
6. C
7. B
8. A
9. C
10. B
11. A
12. A
13. B
14. D
15. C
16. C
17. C
18. C
19. B
20. B
21. B
25. B
26. A
27. C
30. C

MULTI OPTIONS CORRECT

1. A, B, D
2. A, B, D
3. B, C, D
4. A, B, C
5. A, C
7. C, D
9. A, B, C, D
10. B, C
11. A, B
12. A, C, D
13. A, C, D

INTEGER TYPE

1. 1
2. 4
3. 1
4. 1
5. 5
8. 3
9. 6
10. 4
11. $\sqrt{5}$

MATRIX MATCH

1. A – P, B – Q, C – R, D – S
2. A – q, B – p, C – p, D – t
3. A → S, B → P, C → P, D → Q, R

COMPREHENSION for Q 1 - 3

1. A
2. B
3. B

Mathsarc Education

A learning place to fulfill your dream of success!

MATHEMATICS IIT JEE Main/Advanced

STRAIGHT LINES

You cannot walk a straight line without a fixed point to follow.

SUBJECTIVE PROBLEMS (LEVEL - I)

1. Draw the figure of polygon ABCDEF and find its area where points are

   $A\left(\dfrac{5}{2}, 0\right)$, B(1, 1), C(–1, 1), D(–2, 0), E(–1, –1), F(1, –1)

2. Find the distance between the points A & B where points are

   (i) A(0, 0), B(1, 0)　　(ii) A(0,0), B(1, 2)　　(iii) A(1, 1), B(–1, 3)　　(iv) A(a, b), B(b, a)

3. Find the true statements

   (a) In isosceles triangle, at least two angles are equal

   (b) In case of equilateral triangle

      (i) All the sides are equal

      (ii) All angles are 60°

      (iii) Area of triangle $\dfrac{\sqrt{3}}{4}(\text{side})^2$

      (iv) In-centre, Centroid, Orthocenter, Circumcentre are same.

   (c) If $\triangle ABC$ is right angle at B, then $AC^2 = AB^2 + BC^2$ and $\angle B > \angle A$.

   (d) Largest chord of circle subtend 45° at its circumference

   (e) If $\triangle ABC$ is similar to $\triangle DEF$ then $\dfrac{AB}{AC} = \dfrac{DE}{FE}$

4. Match the following

	COLUMN - I		COLUMN - II
A	Square	(p)	Diagonals bisect each other
B	Rhombus	(q)	Diagonal form angle 90°
C	Rectangle	(r)	Opposite side are parallel
D	Parallelogram	(s)	Opposite angles are equal
E	Trapezium	(t)	Sum of opposite angle is 180°
		(u)	Area is product of non-parallel sides
		(v)	Centre of circum circle is point of intersection of diagonal
		(w)	Any line passing through intersection of diagonals divides in two equal part
		(x)	Exactly two sides are parallel
		(y)	Diagonal is an angle bisector
		(z)	In-circle touches all the sides

4. A→all (except x)   B→p, q, r, s, w, y, z, C→p, r, s, t, u, v, w, D→p, r, s, w E→x

5. Find the value of following determinant

(a) $\begin{vmatrix} 2 & 1 \\ 3 & 5 \end{vmatrix}$
(b) $\begin{vmatrix} 1 & 7 \\ 5 & 2 \end{vmatrix}$
(c) $\begin{vmatrix} 1 & 2 & 5 \\ 6 & 7 & 8 \\ 9 & 0 & 7 \end{vmatrix}$
(d) $\begin{vmatrix} 1 & 1 & 0 \\ 0 & 2 & 3 \\ 2 & 5 & 1 \end{vmatrix}$

6. (i) Find the area of Geometry formed by points A, B & C hence identify the geometry.

(A) A(0, 0), B(1,0), C(1,1)

(B) A(0,0), B(2, 0), C(1, $\sqrt{3}$)

(C) A(0,0), B($\sqrt{3}$,1), C (2$\sqrt{3}$,0)

(D) A(1,3), B$\left(\dfrac{1}{2}, 2\right)$, C(−3,−5)

(E) Find the value of x for which the points (x, −1), (2, 1) and (4, 5) are collinear.

7. Prove that the points (a, 0), (0, b) and (1, 1) are collinear if $\dfrac{1}{a}+\dfrac{1}{b}=1$

8. Find area of quadrilateral ABCD where points are as follow: A(1, 1), B(7, –3), C(12, 2) & D(7, 21)

9. Find the coordinates of the points which divide the line segments joining the points (6, 3) and (– 4, 5) in the ratio 3:2

   (i) Internally   (ii) Externally

10. Find the ratio in which y – axis divides the line segment joining the points P(1, 5) & Q(3, 7).

    Ans: 1:3 Externally

11. If a line segment AB, where $A(x_1, y_1)$ and $B(x_2, y_2)$ is divided by a straight line $L \equiv ax + by + c = 0$ in ratio $\lambda : 1$, then $\lambda = -\dfrac{ax_1 + by_1 + c}{ax_2 + by_2 + c}$ or $-\dfrac{L_1}{L_2}$.

12. given that A (1, 1) and B (2, –3) are two points and D is a point on AB produced such that AD = 3AB. Find the coordinates of D.

13. Prove that the points (–2, –1), (1, 0), (4, 3) and (1, 2) are the vertices of a parallelogram. Is it a rectangle?

14. If P (1, 2), Q (4, 6), R (5, 7) and S (a, b) are the vertices of a parallelogram PQRS. Then for parallelogram PQRS

    (A) a = 2, b = 4   (B) a = 3, b = 4   (C) a = 2, b = 3   (D) a = 3, b = 5

15. Find the slope of line passing through points

    (i) A(1,2) & B(2,3)   (ii) A(–2, –1) & B(3, –2)   (iii) A(1, 5) & B(1,6)

    (iv) Making inclination 60° with positive direction of x-axis

16. (a) What is the slope of a line whose inclination is?

    (i) 0°   (ii) 90°   (iii) 120°   (iv) 150°

17. Find the values of p for which (p + 1, 1), (2p + 1, 3) and (2p + 2, 2p) are collinear.

Ans: 2 or – 1/2

18. Line through the point (–2, 6) and (4, 8) is perpendicular to the line through the points (8, 12) and (x, 24). Find the value of x

19. (a) If angle between two lines is $\dfrac{\pi}{4}$ & slope of one of line is 1/2. Find slope of other line.

   (b) If A(- 2, 1), B(2, 3) & C(-2, - 4) are three points. Find angle between BA & BC.

   (c) a ray of light passing through the point (1, 2) reflects on the x-axis at point A and the reflected ray passes through the point (5, 3). Find the co-ordinates of A.

   (d) Find the acute angle between the lines whose slopes are $\sqrt{3}$ & $\dfrac{1}{\sqrt{3}}$.

   Ans: (a) m = 3 or $-\left(\dfrac{1}{3}\right)$ (b) $\theta = \tan^{-1}\left(\dfrac{2}{3}\right)$ (c) (13/5, 0) (d) $\dfrac{\pi}{6}$

20. Find the area of closed curve formed by lines x = 1, x = –1, y = 1 & y = –1.

21. Find the equation of line, if line has following property.

   (a) Passing through (1, 2) and having slope 3   (b) Passing through A(1, 1) & B(3, 2).

   (c) $\tan\theta = \dfrac{1}{2}$ & y - intercept is $-\dfrac{3}{2}$.   (d) $\tan\theta = \dfrac{1}{2}$ & x - intercept = 4.

   (e) Makes intercepts –3 & 4 on the x & y axis respectively.

22. Convert the following lines in various form of line as given below.

   (a) – x + 2y + 1 = 0   (b) 2x + y = 6

      (i) slope–intercept form   (ii) Intercept form

   (b) Find the value of k for which the line $(k - 3)x - (4 - k^2)y + k^2 - 7k + 6 = 0$ is

      (i) Parallel to the x – axis,

      (ii) Parallel to the y-axis,

      (iii) Passing through the origin.

23. Find the equation of the line which passes through the point (3, 4) and the sum

of its intercepts on the axes is 14. **Ans:** x + y = 7 & 4x + 3y = 24

24. Find the slope, x-intercept & y-intercept of the line 3x − 5y − 8 = 0. **Ans:** $\frac{3}{5}, \frac{8}{3}, -\frac{8}{5}$

25. Find the point of intersection of pairs of line as given below

    (a) x + 2y = 1 & 3x + 5y = 2

    (b) x + y = 5 & 2x − y = 6

    (c) 3x + y = 2 & 6x + 2y = 5

26. Prove that the straight lines 4x + 7y = 9, 5x − 8y + 15 = 0 and 9x − y + 6 = 0 are concurrent.

27. Find the value of m so that the lines 3x + y + 2 = 0, 2x − y + 3 = 0 and x + my − 3 = 0 may be concurrent. **Ans:** 4

28. Find the distance of following point from the line 4x + 3y + 1 = 0

    (a) (0,0)

    (b) (1, 2)

    (c) $\left(-\frac{1}{4}, 0\right)$

29. Which one is of largest distance from the line 4x + 3y + 3 = 0

    (a) A(0, 0)
    (b) 4x + 3y + 1 = 0
    (c) 8x + 6y + 3 = 0
    (d) 4x + 3y − 1 = 0

30. Find the distance of a line 2x − y + 3 = 0 from the point (2, 1) measured parallel to the line x + y = 1. **Ans:** 2√2

31. Two sides of a square lie on lines x + y = 1 & x + y + 2 = 0. What is its area? **Ans:** $\frac{9}{2}$

32. Convert the following lines in normal forms

    (a) $\sqrt{3}x + y - 8 = 0$

    (b) $x - \sqrt{3}y + 4 = 0$

33. Find the equation of the perpendicular bisector of the line segment joining the points

    A(2, 3) & B(6, − 5). also find the equation of line perpendicular to AB & passing through a point P on AB such that it divide AB in the ratio 2:1 internally.

34. Let PS be the median of the triangle with vertices P(2,2), Q(6, −1) & R(7, 3). The equation of the line passing through (1, −1) and parallel to PS is?

    (a) 2x − 9y − 7 = 0
    (b) 2x − 9y − 11 = 0
    (c) 2x + 9y − 11 = 0
    (c) 2x + 9y + 7 = 0

35. Which of the following pair of points are same side of line x + 3y - 5 = 0

    (a) (0, 1), (2, −3) (b) (−1, 2), (0, −3) (c) (0, 1), (−5, 0) (d) (0, 0), (1, 2)

36. Plot the region bounded by following

    (i) x ≥ 0 (ii) y < 0 (iii) x − y − 5 < 0 (iv) x ≤ 1

37. Plot the region bounded by the following:

    (i) 2x − y + 1 ≥ 0 (ii) x + y − 2 < 0 (iii) x − axis & y- axis.

38. Are the points (3, − 4) and (2, 6) on the same or opposite side of the line 3x − 4y = 8? **Ans:** Opposite sides

39. Which one of the points (1, 1), (−1, 2) and (2, 3) lies on the side of the line 4x + 3y − 5 = 0, in which the origin lies? **Ans:** (−1, 2)

40. Find the set of values of α such that the point P(0, α) lies inside the triangle formed by the lines x + y − 1 = 0, 2x − y + 1 = 0 and 2x + 6y + 1 = 0. **Ans:** (−1/6, 1)

41. Determine set of values of α for which the point (α, α²) lies inside the triangle formed by the lines 2x + 3y − 1 = 0, x + 2y − 3 = 0 and 5x − 6y − 1 = 0. **Ans:** $\alpha \in \left(-\frac{3}{2}, -1\right) \cup \left(\frac{1}{2}, 1\right)$.

42. In a triangle ABC, E is the midpoint of BC and D is a point on AB such that AD : DB = 2 : 1. If CD and AE intersect at P, determine the ratio CP : PD.

43. The acute angle between two lines is $\frac{\pi}{4}$ and slope of one of them is 1/2. Find the slope of the other line. **Ans:** − 1/3 or 3

44. Find the equation of the straight line which passes through the origin and making angle 60° with the line $x + \sqrt{3}y + 3\sqrt{3} = 0$. **Ans:** x = 0

45. Find the equation of the straight line that passes through the point (3, 4) and perpendicular to the line 3x + 2y + 5 = 0. **Ans:** 2x − 3y + 6 = 0

46. In any triangle ABC, prove that AB² + AC² = 2(AD² + BD²), where D is the midpoint of BC (Apollonius theorem).

47. Find the equation of the straight line which passes through the point (1, 2) and makes an angle θ with the positive direction of the x-axis where cos θ = − 1/3.

48. Find the equation of the line which is at a distance 3 from the origin and the perpendicular from the origin to the line makes an angle of 30° with the positive direction of the x-axis. **Ans:** x + y = 6

49. Find the foot of perpendicular of P (– 3, 5) on the line x – y + 2 = 0. **Ans:** (0,2)

50. Find the image of the point P(–1, 2) in the line mirror 2x – 3y + 4 = 0.

    Ans: $\left(\dfrac{3}{13}, \dfrac{2}{13}\right)$

51. Find the equation of the line which passes through the point (3, 4) and the sum of its intercepts on the axes is 14.

52. If the image of the point (2, 1) with respect to a line mirror be (5, 2), find the equation of the mirror.

53. Show that the points A(4, 4), B(3, 5) and C(–1, –1) form a right angled triangle.

54. If the lines x + 2ay = a, x + 3by = b, x + 4cy + c = 0 are concurrent, then show that a, b, c are in harmonic progression.

55. Two sides of an isosceles triangle are given by the equations 7x – y + 3 = 0 and x + y – 3 = 0 and its third side passes through the point (1, –10). Determine the equation of the third side. **Ans.** x – 3y – 31 = 0 or 3x + y + 7 = 0

56. Find the point P on the line 2x + 3y + 1 = 0, such that |PA – PB| is maximum, where A is (2, 0) and B is (0, 2).

57. Let 2x – 3y = 0 be a given line and P (sinθ, 0) and Q (0, cosθ) be the two points. Then P and Q lie on the same side of the given line, if θ lies in the

    (A) 1st quadrant   (B) 2nd quadrant   (C) 3rd quadrant   (D) none of these

58. The image of the point (3, 8) in the line x + 3y = 7 is

    (A) (1, – 4)   (B) (1, 4)   (C) (-1, – 4)   (D) None of these

LOCUS PROBLEMS

59. The ends of a rod of length ℓ move on two mutually perpendicular lines. Find the locus of the point on the rod, which divides it in the ratio 2: 1. **Ans:** $36x^2 + 9y^2 = 4\ell^2$

60. Find the locus of a point such that the sum of its distances from the points (2, 0) and (- 2, 0) is 6. **Ans:** $\dfrac{x^2}{9} + \dfrac{y^2}{5} = 1$

61. Find the locus of the point of intersection of the lines

    $x\cos\alpha + y\sin\alpha = a$ and $x\sin\alpha - y\cos\alpha = b$, where α is a variable. **Ans:** $x^2 + y^2 = a^2 + b^2$

62. Find the locus of point P such that PA = 3PB where A(0, 0) and B(2, 0).

    Ans: $2x^2 + 2y^2 - 9x + 9 = 0$

63. A(3, 1) and B(2, -5) are the vertices of a triangle. Find the equation of locus of the third vertex C, if the centroid of the triangle lies on the locus $y = 3 + 2x^2$. **Ans:** $2x^2 + 20x - 3y + 89 = 0$.

64. If the coordinates of a variable point P be $(a\cos\theta, b\sin\theta)$ where θ is a variable quantity, find the locus of P. **Ans:** $\dfrac{x^2}{a^2} + \dfrac{y^2}{b^2} = 1$

65. Find the locus of a point P such that the sum of its distances from (0, 2) and (0, -2) is 6.

66. Find the locus of the vertex A of an isosceles triangle whose base is the line joining the points B(2,-3), and C(- 3, 0).

67. Given A(- 5, 2) and B is the point on the locus whose equation is $x^2 + y^2 - 2x + 4y + 8 = 0$. If the point P divides segment AB externally in the ratio 2 : 1, find the equation of locus of P.

    Ans: $x^2 + y^2 - 14x + 12y + 97 = 0$

PARAMETRIC FORM

68. Find the equation of the straight line, which passes through the point (3, 2) and whose slope is 3/4. Find the co-ordinates of the points on the line that are 5 units away from the point (3, 2). **Ans:** $3x - 4y = 1$, P(7, 5), Q(-1, -1)

69. Find the equation of the line through the point A(2, 3) and making an angle of 45° with the x-axis. Also determine the length of intercept on it between A & the line $x + y + 1 = 0$ **Ans:** $x - y + 1 = 0$, $3\sqrt{2}$

70. A straight line is drawn through the point A(1, 0) making an angle of $\dfrac{\pi}{6}$ with positive direction of the x-axis. If it meets the straight line $\sqrt{3}x - 4y + 8 = 0$ in B, find the distance between A and B. **Ans:** $16 + 2\sqrt{3}$

71. Represent the straight-line y = x + 2 in the parametric form.   Ans. $\left(\dfrac{r}{\sqrt{2}},\ 2+\dfrac{r}{\sqrt{2}}\right)$

72. A line joining two points A(2, 0) and B(3, 1) is rotated about A in the anticlockwise direction through an angle of 15°. Find the equation of the line in the new position. If B goes to C, what will be the coordinates of C, in the new position?   Ans. $\left(2+\dfrac{1}{\sqrt{2}},\ \dfrac{\sqrt{3}}{\sqrt{2}}\right)$

73. Show that the equations

    (a) x = 2 + t, y = 3 + 2t and

    (b) x = 3 – 5t, y = 5 – 10t represent the same line.

74. Show that the locus of the mid-point of the distance between the axes of the variable line x cosα + y sinα = p is $\dfrac{1}{x^2}+\dfrac{1}{y^2}=\dfrac{4}{p^2}$ where p is a constant.

75. A line through A(-5, -4) meets the lines x + 3y + 2 = 0, 2x + y + 4 = 0 and x – y – 5 = 0 at the points B, C and D respectively, if $\left(\dfrac{15}{AB}\right)^2+\left(\dfrac{10}{AC}\right)^2=\left(\dfrac{6}{AD}\right)^2$ find the equation of the line.

    Ans: 2x + 3y + 22 = 0

76. Change to cartesian coordinates the equations from given polar form

    (i) r = a sinθ   (ii) $r^{1/2}=a^{1/2}\cos\left(\dfrac{\theta}{2}\right)$

TRIANGLE POINTS

77. Let PS be the median of the triangle with vertices P(2, 2), Q(6, -1) and R(7, 3). The equation of the line passing through (1, -1) and parallel to PS is:

    (A) 4x + 7y + 3 = 0   (B) 2x – 9y – 11 = 0

    (C) 4x – 7y – 11 = 0   (D*) 2x + 9y + 7 = 0

78. The co-ordinates of the incentre of the triangle that has the co-ordinates of mid points of its sides as (0, 1), (1, 1) and (1, 0) is:

    (A*) 2 – √2   (B) 1 + √2   (C) 1 – √2   (D) 2 + √2

79. Consider the triangle OAB, where 'O' is the origin. If B ≡ (3, 4) and orthocenter of the triangle is P ≡ (1, 4), find the coordinates of A.

80. Find the circumcenter of ΔABC where A(4, -3), B(-2, 1) and C(2, 3). **Ans:** $\left(\dfrac{9}{7}, -\dfrac{4}{7}\right)$

81. Find the circumcenter of triangle having sides: x = 2, y + 1 = 0 and x + 2y = 4.
    Ans: (4, 0)

82. The vertices of a triangle are A(10, 4), B(-4, 9) and C(-2, -1). Find the equation of its altitudes. Also find its orthocenter.

    Ans. x - 5y + 10 = 0, 12x + 5y + 3 = 0, 14x - 5y + 23 = 0, $\left(-1, \dfrac{9}{5}\right)$.

83. Let ABC be a triangle with equations of the sides AB, BC and CA respectively x - 2 = 0, y - 5 = 0 and 5x + 2y - 10 = 0. Then the orthocenter of the triangle lies on the line

    (A) x - y = 0   (B) 3x - y = 1   (C) x - 2y = 1   (D) none of these

    Ans: B, Orthocenter is (2, 5)

84. Find the values of non-negative real numbers h_1, h_2, h_3, K_1, K_2, K_3 such that the algebraic sum of the perpendiculars drawn from points (2, k_1), (3, k_2), (7, k_3), (h_1, 4) (h_2, 5), (h_3, -3) on a variable line passing through (2, 1) is zero.

85. If ΔABC has orthocenter (1, 1) & circumcenter $\left(\dfrac{3}{2}, \dfrac{3}{4}\right)$ then find its centroid. **Ans:** $\left(\dfrac{4}{3}, \dfrac{5}{6}\right)$

ANGLE BISECTORS + FAMILY OF LINES

86. Obtain the equations of the lines passing through the intersection of lines 4x - 3y - 1 = 0 and 2x - 5y + 3 = 0 and equally inclined to the axes.

    Ans: x + y - 2 = 0 and x = y

87. Find the equation of line through the intersection of given lines $\begin{cases} x+y-1=0 \\ 2x-y+3=0 \end{cases}$, satisfying the given condition.

    (a) through (1, 2)   **Ans:** x - 5y + 9 = 0

    (b) parallel to 2x + 3y = 1   **Ans:** 6x + 9y - 11 = 0

(c) perpendicular to 2x + 3y = 1   **Ans:** 9x – 6y + 16 = 0

88. Show that the family of lines (2 – λ)x + (3λ +1)y + 6 = 0, where λ ∈ R, passes through the fixed point. Also find the fixed point.   **Ans:** $\left(-\dfrac{18}{7}, -\dfrac{6}{7}\right)$

89. Find the equations of the bisectors of the angle between the straight lines 3x – 4y + 7 = 0 and 12x – 5y – 8 = 0.

    Ans: 21x + 27y – 131 = 0, 99x – 77y + 51 = 0

90. Find the equations of the bisectors of the angles between the following pairs of straight lines 3x + 4y + 13 = 0 and 12x – 5y + 32 = 0.

    Ans: 21x – 77y – 9 = 0 & 99x + 27y + 329 = 0

91. For straight lines 4x + 3y – 6 = 0 and 5x + 12y + 9 = 0, find the equation of the

    (i) Bisector of the obtuse angle between them;

    (ii) Bisector of the acute angle between them;

    (iii) Bisector of the angle which contains (1, 2);

    (iv) Bisector of the angle which contains the origin.

    Ans: 9x – 7y – 41 = 0, 7x + 9y – 3 = 0, 9x – 7y – 41 = 0, 7x + 9y – 3 = 0

SUBJECTIVE PROBLEMS (LEVEL - II)

1. A ray of light coming from the point (1, 2) is reflected at a point A on x-axis and then passes through the point (5, 3). The coordinates of the point A are   **Ans.** $\left(\dfrac{13}{5}, 0\right)$

2. The opposite angular points of a square are (3, 4) and (1, –1) then the coordinates of the other two vertices are   **Ans.** $\left(-\dfrac{1}{2}, \dfrac{5}{2}\right), \left(\dfrac{9}{2}, \dfrac{1}{2}\right)$

3. A vertex of an equilateral triangle is (2, 3) and the opposite side is x + y = 2. The equations of other sides are?   **Ans.** $(2+\sqrt{3})x - y = 1 + 2\sqrt{3}$; $(2-\sqrt{3})x - y = 1 - 2\sqrt{3}$

4. If the two pairs of lines $x^2 - 2mxy + y^2 = 0$ and $x^2 - 2nxy - y^2 = 0$ one such that one of them represents the bisector of angle between the others, then: **Ans.** mn + 1 = 0

5. A line passes through the point P(2, 3) and makes an angle θ with positive direction of x-axis. If it meets the lines represented by x²–2xy –y²=0 at the points A and B. If PA.PB = 17, if $\theta \in [0, \pi]$ then number of such θ is equal to **Ans.** 2

6. Let P = (1, 1) and Q = (3, 2). The point R on the x-axis such that PR + RQ is minimum is given by $\left(\dfrac{\lambda}{3}, 0\right)$, where λ = **Ans.** 5

7. Consider a family of straight lines (x + y) + λ(2x–y+1) = 0. Find the equation of the straight line belonging to this family that is farthest from (1, –3). **Ans.** 15y – 6x – 7=0

8. If the quadratic equation $ax^2 + bx + c = 0$ has –2 as one of its roots then ax + by + c = 0 represents **Ans.** A family of concurrent lines

9. The line L has intercepts a & b on the co-ordinate axes. Keeping the origin fixed the co-ordinate axes are rotated through a fixed angle. The line L has now intercepted p and q on the rotated axes. Find the relation in terms of a, b, p & q. **Ans.** $\dfrac{1}{a^2} + \dfrac{1}{b^2} = \dfrac{1}{p^2} + \dfrac{1}{q^2}$

10. Two sides of a rhombus OABC (lying entirely in first quadrant or fourth quadrant) of area equal to 2 sq. units, are $y = \dfrac{x}{\sqrt{3}}$, $y = \sqrt{3}\,x$. Then possible coordinates of B is / are ('O' being the origin) **Ans.** $\left(1+\sqrt{3},\ 1+\sqrt{3}\right)$

Mathsarc Education

A learning place to fulfill your dream of success!

MATHEMATICS IIT JEE Main/Advanced

LIMITS
You are your only limit.

SINGLE OPTION CORRECT

1. The graph of the function y = f(x) has a unique tangent at $(e^a, 0)$ through which the graph passes then $\lim\limits_{x \to e^a} \dfrac{\ln(1+7f(x)) - \sin(f(x))}{3f(x)}$ is equal to

   (A) 1 (B) 2 (C) 3 (D) 4

2. Evaluate the following limits:

   (i) $\lim\limits_{x \to 0} \dfrac{e^x \sin x - x - x^2}{x^2 + x - \ln(1-x)}$

   (ii) $\lim\limits_{x \to 1} \dfrac{x^x - x}{x - 1 - \ln(x)}$

   (iii) $\lim\limits_{x \to 0} \dfrac{x^{6000} - (\sin x)^{6000}}{x^2 \sin^{6000}(x)}$

   (iv) $\lim\limits_{x \to 0} \dfrac{1 - \cos x \cdot \cos 2x \cdot \cos 3x}{x^2}$

   (v) $\lim\limits_{n \to \infty}(1+x)(1+x^2)(1+x^4)\ldots(1+x^{2n})$, where $|x|<1$

   (vi) $\lim\limits_{n \to \infty} \sqrt[3]{n^2 - n^3} + n$

   (vii) $\lim\limits_{n \to \infty} \dfrac{(n!)^{1/n}}{n}$

   (viii) $\lim\limits_{x \to 0} \dfrac{x}{a}\left[\dfrac{b}{x}\right]$ where [.] = GIF

   (ix) $\lim\limits_{x \to 0} \dfrac{(1+x)^{1/x} - e}{x}$

   (x) $\lim\limits_{x \to 0}\left[\dfrac{a \sin x}{x}\right] + \left[\dfrac{b \tan x}{x}\right]$ where [.] = GIF and $a, b \in I^+$

   (xi) for $x > 0$, $\lim\limits_{x \to 0}\left((\sin x)^{1/x} + \left(\dfrac{1}{x}\right)^{\sin x}\right)$

   (xii) $\lim\limits_{n \to \infty} \sum\limits_{r=1}^{n}\left(\dfrac{r}{1+r^2+r^4}\right)$

(xiii) $\lim_{n\to\infty} n(n\{\ln n - \ln(n+1)\}+1)$

(xiv) $\lim_{x\to\infty}\left(\dfrac{1^{1/x}+2^{1/x}+3^{1/x}+..+n^{1/x}}{n}\right)^{nx}$, $n \in N$

(xv) $\lim_{n\to\infty} \cos^n\left(\dfrac{x}{\sqrt{n}}\right)$

(xvi) $\lim_{n\to\infty}\left(\dfrac{a-1+\sqrt[n]{b}}{a}\right)^n$, where $a>0, b>0, n\in N$

(xvii) $\lim_{n\to\infty}\left[n\tan(n!\,4\pi e)\right]$

(xviii) $\lim_{n\to\infty}\left(1+\sum_{k=1}^{n}\dfrac{2}{{}^nC_k}\right)$

(xix) $\lim_{x\to\infty}\left(\dfrac{e^{\pi/x}+e^{-\pi/x}}{2\cos(\pi/x)}\right)^{x^2}$

(xx) $\lim_{x\to 0}\left(\left[\dfrac{\sin x}{x}\right]+\left[\dfrac{2\sin 2x}{x}\right]+\left[\dfrac{3\sin 3x}{x}\right]+.....+\left[\dfrac{10\sin 10x}{x}\right]\right)$ where $[.] = $ GIF

(xxi) Evaluate: $\lim_{x\to\infty} n^{-n^2}\left\{(n+2^0)(n+2^{-1})(n+2^{-2})(n+2^{-3})......(n+2^{-n+1})\right\}^n$

(xxii) Show that $\lim_{n\to\infty}\sum_{k=0}^{n}\dfrac{{}^nC_k}{n^k(k+3)}=e-2$

(xxiii) $\lim_{n\to\infty} n^2\left(x^{1/n}-x^{1/(n+1)}\right)=$ ___ , $x>0$

(xxiv) $\lim_{x\to 1}\left[\csc\dfrac{\pi x}{2}\right]^{1/(1-x)}=$ ___ , $[.]=$ GIF

3. Let $l_1=\lim_{x\to 0}\dfrac{e^{x\tan(20007)x}-1}{x\log(1+x)}$ & $l_2=\lim_{x\to\infty} x\left(2x-\sqrt[3]{x^3+x^2+1}-\sqrt[3]{x^3-x^2+1}\right)$, then

(A) $I_1 \times I_2 = 4446$ (B) $I_1 \times I_2 = 2223$ (C) $I_1 \times I_2 = 1$ (D) None of these

4. Prove that $\lim_{m\to\infty}\lim_{n\to\infty}\left(1+\cos^{2m}(n!\,\pi x)\right)=\begin{cases}2, & \text{if } x\in Q\\ 1, & \text{if } x\notin Q\end{cases}$.

5. If $\lim_{x\to\infty}\sqrt{(1-x+x^2)(1+x+x^2)}-x^2+b=a$, then

(A) $a = 1, b = -2$ (B) $a = 1, b = 1$ (C) $a = 0, b = -1/2$ (D) $a = 0, b = 1/2$

6. The integer n for which $\lim\limits_{x \to 0} \dfrac{(\cos x - 1)(\cos x - e^x)}{x^n}$ = finite non-zero number is

   (A) 1 (B) 2 (C) 3 (D) 4

Paragraph Q. No. 7 – 8

Let $f(x) = ax^2 + x + 3$ & $f(x) \geq 0 \; \forall \, x \in R, \; \forall \, a \in A$ where $A \subset R$. Also $L = \lim\limits_{x \to \infty}\left(x + 1 - \sqrt{ax^2 + x + 3}\right)$.

7. Range of a is equal to

   (A) $(0, 1)$ (B) $[1, \infty)$ (C) $\left[\dfrac{1}{12}, \infty\right)$ (D) $\left(-\infty, \dfrac{1}{12}\right]$

8. Which one of the following statements is incorrect?

   (A) If L exist then $a = 1$.

   (B) If L does not exist then range of a is $\left[\dfrac{1}{12}, 1\right) \cup (1, \infty)$.

   (C) $|f(x)|$ is continuous and differentiable $\forall \, x \in R, \; \forall \, a \in A$.

   (D) $f(|x|)$ is non-derivable at exactly two points.

9. The value of $\lim\limits_{x \to \infty} \dfrac{x^p + x^{p-1} + 1}{x^q + x^{q-2} + 2}$, where $p > 0, q > 0$ is

   (A) 0 if $p < q$ (B) 1 if $p = q$ (C) ∞, if $p > q$ (D) 1 if $p > q$

10. Select the correct statement ([.] = G.I.F.)

    (A) $[0.\overline{9}] = 1$
    (B) $\lim\limits_{x \to 1^-}[x] = 0$

    (C) $\lim\limits_{x \to 0^+} x^{2x} = 0$
    (D) $\lim\limits_{x \to 0^+} \sqrt{x + \sqrt{x + \sqrt{x + \ldots \infty}}} = 1$

11. Consider the function f(x) as shown in figure and select the correct options.

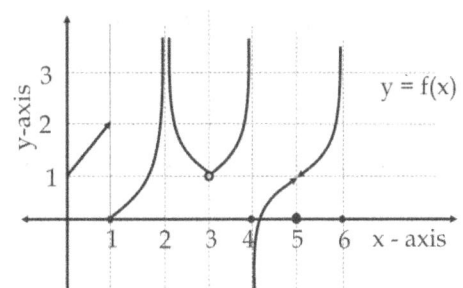

(A) $\lim\limits_{x \to 1} f(x) = D.N.E.$

(B) $\lim\limits_{x \to 2} f(x) = D.N.E.$

(C) $\lim\limits_{x \to 4^-} f(x) = \infty$

(D) $\lim\limits_{x \to 3} f(x) = 1$

12. The value of a for which $\lim\limits_{x \to 0} \dfrac{(e^x - 1)^4}{\sin\left(\dfrac{x^2}{a^2}\right) \log_e\left(1 + \dfrac{x^2}{2}\right)} = 8$ is

(A) -2 (B) -1 (C) 1 (D) 2

13. The largest value of the non-negative integer a for which $\lim\limits_{x \to 1}\left\{\dfrac{-a + \sin(x-1) + a}{x + \sin(x-1) - 1}\right\}^{\frac{1-x}{1-\sqrt{x}}} = \dfrac{1}{4}$, is ___

14. Let f(x) = $\dfrac{e^x \cos x - \ln(1+x) - 1}{x}$, $x \neq 0$ then $\lim\limits_{x \to 0} f(x)$ is equal to ___

15. If $\lim\limits_{x \to \infty}\left(\sqrt{x^4 + ax^3 + 3x^2 + bx + 2} - \sqrt{x^4 + 2x^3 - cx^2 + 3x - d}\right)$ = finite, then the value of a is ___

16. The value of $\lim\limits_{x \to \infty} \dfrac{x}{x + \dfrac{\sqrt[3]{x}}{x + \dfrac{\sqrt[3]{x}}{x + ... \infty}}}$ is equal to ___

17. Let α, β ∈ R is such that $\lim\limits_{x \to 0} \dfrac{x^2 \tan(\alpha x)}{\beta x - \tan(2x)} = 1$, then the value of 5β + 3α, is ___

18. Consider $f(x) = x\sin\left([x]^4 - 5[x]^2 + 4\right)$, then number of points in (-5, 5) where $\lim\limits_{x \to a} f(x) = D.N.E.$ & $a \in (-5, 5)$? Where [.] = GIF.

19. The value of $\left|\lim\limits_{x\to\frac{3\pi}{4}}\left(\dfrac{4\sin^2 x\cos x-\cos x+\sin x}{\sin x+\cos x}\right)\right|$ is equal to _____ where |.| = Absolute value function.

20. Let m, n be two positive integers greater than 1. If $\lim\limits_{\alpha\to 0}\left(\dfrac{e^{\cos(\alpha^n)}-e}{\alpha^m}\right)=-\dfrac{e}{2}$, then value of $\dfrac{m}{n}$ is ___

21. The integral value of n so that $\lim\limits_{x\to 0}\dfrac{(\sin x-x)\left(2\sin x-\ln\left(\dfrac{1+x}{1-x}\right)\right)}{x^n}=finite$ non zero number is _____

22. for $a\in I$, $a\ne -1$, $\lim\limits_{n\to\infty}\dfrac{1^a+2^a+3^a+\ldots+n^a}{(n+1)^{a-1}\left[(na+1)+(na+2)+(na+3)+\ldots+(na+n)\right]}=\dfrac{1}{60}$ then value of a is __

23. 13. Let $f(x)=\cos 2x\cdot\cos 4x\cdot\cos 6x\cdot\cos 8x\cdot\cos 10x$ then $\lim\limits_{x\to 0}\dfrac{1-f^3(x)}{5\sin^2 x}$ equals

   (A) 660   (B) 135   (C) 132   (D) 66

24. Evaluate the limit: $L=\lim\limits_{x\to\infty}x^p\left(\sqrt[3]{x+1}+\sqrt[3]{x-1}-2\sqrt[3]{x}\right)=$ Exist and non-zero.

25. If $\lim\limits_{x\to 0}\left(x^{-3}\sin 3x+ax^{-2}+b\right)$ exist and is equal to zero then:

   (A) a = - 3 & b = 9/2   (B) a = 3 & b = 9/2
   (C) a = - 3 & b = - 9/2   (D) a = 3 & b = - 9/2

26. The value of $\lim\limits_{x\to\infty}\dfrac{\left(2^{x^n}\right)^{1/e^x}-\left(3^{x^n}\right)^{1/e^x}}{x^n}$ (where $n\in N$) is

   (A) $\ln\left(\dfrac{2}{3}\right)$ (B) 0   (C) $n\ln\left(\dfrac{2}{3}\right)$ (D) not defined

27. The value of the $\lim\limits_{t\to 0}\dfrac{\ln(\cos(\sin t))}{t^2}$ is

   (A) - 1/2   (B) 1/2   (C) 1   (D) - 1

28. Which of the following limits is/are equal to unity?

(A) $\lim_{x\to 0^+}\left((\tan x)^{\sin x}+(\sin x)^{\tan x}\right)$

(B) $\lim_{x\to 0^+}\dfrac{\sqrt{1-e^{-x}}-\sqrt{1-\cos x}}{\sqrt{\sin x}}$

(C) $\lim_{x\to 0}\left([1+|x|]\right)^{\frac{1}{|x|}}$ where [.] = GIF

(D) $\lim_{n\to\infty}\dfrac{\sum_{k=1}^{n+1}k^2}{\sum_{k=1}^{n}k^2},\ n\in N$

29. Let $f(x)=\underset{n\to\infty}{Lt}\ln\left(\sqrt{e^{\cos x}\sqrt{e^{3\cos x}\sqrt{e^{5\cos x}......\sqrt{e^{(2n+1)\cos x}}}}}\right)$ and $g(x)=\left[\dfrac{f(x)}{3}\right]$ then g(0) = ___ (where [.] = GIF).

30. $\lim_{n\to\infty}\left[\dfrac{e^{1/n}+2e^{2/n}+3e^{3/n}+.....+ne}{n^2}\right]$ is less than –

(A) 0   (B) 2   (C) 5   (D) 8

31. $\lim_{n\to\infty}\dfrac{1}{n^3}\left[1^2\sin\left(\dfrac{1}{n}\right)+2^2\sin\left(\dfrac{2}{n}\right)+3^2\sin\left(\dfrac{3}{n}\right)+.....+n^2\sin 1\right]$ equals –

(A) cos1 + 2sin1   (B) 2sin1 – 2   (C) cos1 – 2sin1 – 2   (D) cos1 + 2sin1 – 2

32. Let $L=\lim_{n\to\infty}n^2\left(\dfrac{\tan^{-1}(n+1)}{n+1}-\dfrac{\tan^{-1}(n)}{n}\right)-\dfrac{\pi}{2}$, then [- L] is equal to _____ (where [.] = GIF).

33. If $\lim_{x\to\infty}\left(1+\dfrac{1}{x(x+2)}\right)^{x^3}\left(1+\dfrac{1}{x(x+4)}\right)^{-x^3}$ is equal to e^n, then n is equal to ____

34. If $\lim_{n\to\infty}\dfrac{n\cdot 3^n}{n(x-2)^n+n\cdot 3^{n+1}-3^n}=\dfrac{1}{3}$, then the range of x is (where n ∈ N)

(A) [2, 5)   (B) (1, 5)   (C) (-1, 5)   (D) R

THANKS!

Keep smiling!

Visit Us: https://www.mathsarc.com

ANSWER KEY - Limits

1. B
2. (i) 0   (ii) 2   (iii) 1000   (iv) 7

   (v) $\dfrac{1}{1-x}$ (vi) 1/3   (vii) 1/e   (viii) b/a

   (ix) - e/2   (x) a + b – 1   (xi) 1   (xii) 1/2

   (xiii) 1/2   (xiv) n!   (xv) $e^{-(x^2/2)}$ (xvi) $b^{1/a}$

   (xvii) 12   (xviii) e^2 (xix) e^{π^2} (xx) 375

   (xxi) e^2 (xxiii) lnx   (xxiv) 1

3. A   5. C   6. C   7. C
8. D   9. A, B, C   10. A, B, D   11. A, B, C, D.
12. A, D   13. 2   14. 0   15. 2
16. 1   17. 2   18. 5   19. 1
20. 2   21. 6   22. 7   23. D
24. - 2/9   25. A   26. B   27. A
28. B, C, D   29. 1   30. B, C, D   31. D
32. 3   33. 2   34. C

Mathsarc Education

A learning place to fulfill your dream of success!

MATHEMATICS   IIT JEE Main/Advanced

DIFFERENTIABILITY

Ability to differentiate along with smoothness of curves.

CLASS WORK - UNDERSTANDING

1. If $f'(a^+) = \lim\limits_{x \to a^+} \dfrac{f(x) - f(a)}{x - a}$ & $f'(a^-) = \lim\limits_{x \to a^-} \dfrac{f(x) - f(a)}{x - a}$ then find $f'(a^+)$ & $f'(a^-)$ for the following functions and comment about existence of $f'(a)$.

   (i) $f(x) = 2x^2 - x + 3$, where $a = 2$

   (ii) $f(x) = |\ln(x)|$, where $a = 1$

   (iii) $f(x) = e^{-|x|}$, where $a = 0$

   (iv) $f(x) = \cos x$, where $a = 0$

   (v) $f(x) = \begin{cases} x^2 + 1, & x \geq 0 \\ -x^2, & x < 0 \end{cases}$, where $a = 0$

   (vi) $f(x) = \begin{cases} (x - e) \cdot 2^{-2/(e-x)} & x \neq e \\ 0, & x = e \end{cases}$, where $a = e$

2. The slope of the curve $y = f(x)$ at the point $P(x_0, f(x_0))$ is the number
   $$m = \lim_{h \to 0} \dfrac{f(x_0 + h) - f(x_0)}{h} = \text{Exist}.$$

   The tangent line to the curve at P is the line through P with given slope m.

   Now, answer the following question based on above theory.

   (i) Find the slope of the curve $y = 1/x$ at any point $x = a \neq 0$. What is the slope at the point $x = -1$?

   (ii) Where does the slope equals $-1/4$?

   (iii) Write equation of tangent at $P(2, 1/2)$.

3. Consider $\nabla y = \dfrac{\Delta y}{\Delta x}$ or gradient $= \dfrac{y_2 - y_1}{x_2 - x_1}$, where $y = f(x)$. Select the correct options.

   (A) $\lim\limits_{\Delta x \to 0} \dfrac{\Delta y}{\Delta x} = \dfrac{dy}{dx}$.

   (B) $f'(x_1) = \lim\limits_{x \to x_1} \dfrac{y - y_1}{x - x_1}$ and point $P(x_1, y_1)$ lies on the curve $y = f(x)$.

   (C) Greater gradient implies greater rate of change.

   (D) Equation of tangent at point $P(x_1, y_1)$ lies on the curve $y = f(x)$ is: $y - y_1 = \dfrac{dy}{dx}\bigg|_{atP} (x - x_1)$.

4. Explain the geometrical meaning of f'(a) and discuss its different aspect of differentiability of f(x) at x = a.

5. Differentiate the following functions using first principal.

   (i) $f(x) = \dfrac{x}{x-1}$

   (ii) $f(x) = \sqrt{x}$ for $x > 0$.

   (iii) $f(x) = x^n$, $n \in Q$.

   (iv) $f(x) = \ln(1 + x)$

   (v) $\tan(\sqrt{x})$

   (vi) $f(x) = \tan^{-1}(x)$

   (vii) $f(x) = e^{\sqrt{3x+2}}$

   (viii) $f(x) = \ln(3x + 2)$, also find f'(0).

6. When does f(x) said to be non-differentiable at x = a (finite). Select the correct options and Justify using an appropriate Example?

   (A) If $f'(a^+) \neq f'(a^-)$ (both finite).

   (B) If $f'(a^+) = \infty$ & $f'(a^-) = -\infty$ or $f'(a^+) = -\infty$ & $f'(a^-) = \infty$.

   (C) If f(x) is discontinuous at x = a.

   (D) f(x) has vertical tangent at x=a. Ex $f(x) = x^{1/3}$ at x=0

7. (i) Let f(x) be a function satisfying $|f(x)| \leq x^2$ for $-1 \leq x \leq 1$. Show that f is differentiable at x = 0 and find f'(0).

   (ii) Show that $f(x) = \begin{cases} x^2 \sin\left(\dfrac{1}{x}\right), & x \neq 0 \\ 0, & x = 0 \end{cases}$ is differentiable at x = 0 and find f'(0).

8. Prove that differentiability of f(x) at x = a implies continuity at x = a but converse is not true.

9. Explain the differentiability of f(x) in x ∈ (a, b) or x ∈ [a, b] and check the differentiability of
$$f(x) = \begin{cases} |1-4x^2|, & 0 \leq x < 1 \\ [x^2 - 2x], & 1 \leq x \leq 2 \end{cases}$$ in x ∈ (0, 2). Where [.] = GIF & |.| = Modulus function.

10. If $f(x) = \begin{cases} ax + b, & x \leq -1 \\ ax^3 + x + 2b, & x > -1 \end{cases}$ is differentiable for all x ∈ R. Find 'a' & 'b'.

11. If $f(x) = \begin{cases} x^m \cdot \sin\left(\dfrac{1}{x}\right) & x \neq 0 \\ 0 & x = 0 \end{cases}$ is continuous but not differentiable at x = 0, then find m.

12. If $f(x) = \begin{cases} \sqrt{4x^2 - 12x + 9} \cdot \{x\}, & x \geq 1 \\ \cos\left(\dfrac{\pi(|x| - \{x\})}{2}\right), & x < 1 \end{cases}$ then check the differentiability in [-1, 2].

13. Prove that f'(x) = u(x)·v'(x) + u'(x)·v(x) where f(x) = u(x)·v(x).

14. Let f(x) = 15 - |x - 10|; x ∈ R. Then the set of all values of x, at which the function, g(x) = f(f(x)) is not differentiable, is:

    (A) {5, 10, 15} (B) {10, 15} (C) {5, 10, 15, 20} (D) {10}

15. Let S = {t ∈ R: f(x) = |x - π|($e^{|x|}$-1)sin|x| is not differentiable at t}. Then the set S is equal to

    (A) {0} (B) {π} (C) {0, π} (D) ∅ (an Empty set)

16. For x ∈ R, f(x) = |log2 - sinx| and g(x) = f(f(x)), then:

    (A) g'(0) = - cos(log2)

    (B) g is differentiable at x = 0 and g'(0) = - sin(log2).

    (C) g is not differentiable at x = 0

    (D) g'(0) = cos(log2)

17. If the function $g(x) = \begin{cases} k\sqrt{x+1}, & 0 \leq x \leq 3 \\ mx + 2, & 3 < x \leq 5 \end{cases}$ is differentiable, then the value of k + m is

    (A) 10/3 (B) 4 (C) 2 (D) 16/5

18. Let f:R→R and g:R→R be respectively given by f(x) = |x| + 1 and g(x) = x^2 + 1. Define h:R →R by $h(x) = \begin{cases} \max\{f(x), g(x)\} & \text{if } x \leq 0 \\ \min\{f(x), g(x)\} & \text{if } x > 0 \end{cases}$. The number of points at which h(x) is not differentiable is ____

19. If $|c| \leq \frac{1}{2}$ and f(x) is a differentiable function at x = 0 given by $f(x) = \begin{cases} b\sin^{-1}\left(\frac{c+x}{2}\right), & -\frac{1}{2} < x < 0 \\ 1/2, & x = 0 \\ \frac{e^{ax/2} - 1}{x}, & 0 < x < \frac{1}{2} \end{cases}$.

    Find the value of 'a' and prove that 64 b^2 = 4 – c^2.

20. The left hand derivative of f(x) = [x]sin(πx) at x = k, k an integer, is

    (A) $(-1)^k$ (k - 1)π   (B) $(-1)^{k-1}$(k - 1)π   (C) $(-1)^k$ kπ   (D) $(-1)^{k-1}$ kπ

FUNCTIONAL RELATIONSHIP

1. Let f be a differentiable function satisfying $f\left(\frac{x}{y}\right) = f(x) - f(y)$ for all x, y > 0. If f'(1) = 1 then find f(x).

2. A differentiable function satisfying the relation $f(x+y) = f(x) + f(y) + 2xy - 1 \, \forall x, y \in R$. If $f'(0) = \sqrt{3 + a - a^2}$ find f(x) and prove that f(x) > 0 ∀ x ∈ R.

3. If f(x + y) = f(x)·f(y), ∀ x, y ∈ R then prove that f(kx) = f(x)^k for ∀ k, x ∈ R.

4. Let f:R→(- π, π) be differentiable function such that $f(x) + f(y) = f\left(\frac{x+y}{1-xy}\right)$. If $f(1) = \frac{\pi}{2}$ and $\lim_{x \to 0} \frac{f(x)}{x} = 2$, find f(x).

THANKS!

Keep smiling!

Visit Us: https://www.mathsarc.com

ANSWER KEY - Differentiability

1. (i) Exist, f'(2) = 7 (ii) DNE, f'(1^+) = 1, f'(1^-) = -1
    (iii) DNE, f'(0^+) = -1, f'(0^-) = 1 (iv) Exist, f'(0) = 0
    (v) DNE, f'(0^+) = 0, f'(0^-) = ∞ (vi) DNE, $f'(e^+) = \infty, f'(e^-) = 0$

2. (i) $-1/a^2$, -1 (ii) a = ± 2 (iii) x + 4y – 4 = 0

3. A, B, C

5. (i) $f'(x) = -\dfrac{1}{(x-1)^2}$ (ii) $f'(x) = \dfrac{1}{2\sqrt{x}}$ (iii) $f'(x) = n \cdot x^{n-1}$

    (iv) $f'(x) = \dfrac{1}{1+x}$ (v) $f'(x) = \sec^2(\sqrt{x}) \cdot \dfrac{1}{2\sqrt{x}}$ (vi) $f'(x) = \dfrac{1}{1+x^2}$

    (vii) $f'(x) = e^{\sqrt{3x+2}} \times \dfrac{1}{2\sqrt{3x+2}} \times 3$ (viii) $f'(x) = \dfrac{3}{3x+2}, f'(0) = \dfrac{3}{2}$

6. A, B, C, D   7. (i) 0 (ii) 0   9. At x = 1/2, 1   10. a = -1/2, b = 1

11. m ∈ (0, 1]   12. Non-differentiability at x = 0, 1, 3/2, 2   14. A

15. D   16. D   17. C   18. 3

19. a = 1   20. A

FUNCTIONAL RELATIONSHIP

1. f(x) = ln(x)   2. $f(x) = x^2 + \left(\sqrt{3+a-a^2}\right)x + 1$ 3. $f(x) = e^{\lambda x}$

4. $f(x) = 2\tan^{-1} x$

Mathsarc Education

A learning place to fulfill your dream of success!

MATHEMATICS   IIT JEE Main/Advanced

DIFFERENTIATION - BASIC

A big goal can be achieved easily through differentiation.

CLASS WORK - UNDERSTANDING

1. Find $\dfrac{dy}{dx}$ for the following

   (a) $y = x^3 + 2x + 3$

   (b) $y = kx^2 + c$ where k & c are any two real constant

   (c) $y = x^2 + \sin x$

   (d) $y = x + \dfrac{1}{x} + \ln x + 3^x$

   (e) $y = 2\cos x + 3\sin x + \tan x$

   (f) $y = tx^2 + 1$ where t is independent on x

   (g) $y = xt + 2t + x$, find $\dfrac{dy}{dt}$

   (h) $y = \sec x + \tan x + x^{\frac{1}{3}} + \dfrac{1}{x} + x^{\frac{1}{2}}$

   (i) $y = tx^2 + 1$ if t is dependent on x such that $t = (2x - 1)$

2. Find $\dfrac{dy}{dx}$ for the following

   (a) $y = xe^x$

   (b) $y = x^2 \sin x + x \ln x$

   (c) $y = x^n \cos x + e^x \sin x$

   (d) $y = t \sin x$ if t is independent on x

   (e) $y = t \sin x$ if $t = xe^x$

3. Find $\dfrac{dy}{dx}$ for the following

   (a) $y = \dfrac{x}{1+x}$

   (b) $y = \dfrac{e^x}{1+x}$

   (c) $y = \dfrac{e^x + 1}{1 + \sin x}$

(d) $y = \dfrac{t}{t+x}$ if t is independent on x

(e) $y = \dfrac{t}{t+x}$ if $t = e^x$

4. Find $\dfrac{dy}{dx}$ if

   (a) $y = \sin(x + y)$

   (b) $y = \tan\left(e^{(x+y)}\right)$

5. If $x = a(\theta + \sin\theta)$, $y = a(1 - \cos\theta)$, Find $\dfrac{dy}{dx}$,   Ans. $\tan\theta/2$

6. If $x = a\cos^3\theta$, $y = a\sin^3\theta$, Find $\dfrac{dy}{dx}$ Ans. $-\tan\theta$

7. If $x = \log t + \sin t$, $y = e^t + \cos t$, find $\dfrac{dy}{dx}$,   Ans. $\dfrac{t(e^t - \sin t)}{1 + t\cos t}$

8. If $y = x^x$, Find $\dfrac{dy}{dx}$ Ans. $x^x(1 + \log x)$

9. If $y = (\sin x)^{\cos x} + (\cos x)^{\sin x}$,

   Prove that $\dfrac{dy}{dx} = (\sin x)^{\cos x}\left\{\cot x \cdot \cos x - \log(\sin x)^{\sin x}\right\} + (\cos x)^{\sin x}\left\{\log(\cos x)^{\cos x} - \tan x \cdot \sin x\right\}$

10. If $y = \cos\sqrt{\sin\sqrt{x}}$, Find $\dfrac{dy}{dx}$

11. If $xy = x^3 + y^3$, find $\dfrac{dy}{dx}$

12. If $x + y = \sin(xy)$, Find $\dfrac{dy}{dx}$

13. If $y = \tan(x + y)$, Find $\dfrac{dy}{dx}$

14. If $x^3 + y^3 = \sin(x + y)$, Find $\dfrac{dy}{dx}$

15. If $x = y\log(xy)$, Find $\dfrac{dy}{dx}$

16. If $\sin y = x\sin(a + y)$, Prove That $\dfrac{dy}{dx} = \dfrac{\sin^2(a+y)}{\sin a}$

17. If $y = \sqrt{\sin x + \sqrt{\sin x + \sqrt{\sin x + \ldots \infty}}}$. Prove that $\dfrac{dy}{dx} = \dfrac{\cos x}{2y - 1}$

18. If $x = y + \cfrac{1}{y + \cfrac{1}{y + \cfrac{1}{y + \ldots \infty}}}$, Prove that $\dfrac{dy}{dx} = 2x^2 + y^2 - 3xy$

19. If $\cos y = x\cos(b + y)$, prove that $\dfrac{dy}{dx} = \dfrac{\cos^2(b+y)}{\sin b}$

20. If $y = \dfrac{e^x}{\log x}$, Find $\dfrac{dy}{dx}$

21. If $y = \sin(\cot x)$, Find $\dfrac{dy}{dx}$

22. If $y = \sin(\sqrt{\cos x})$, Find $\dfrac{dy}{dx}$

23. If $y = \sqrt{\sin \sqrt{x}}$, Find $\dfrac{dy}{dx}$

SOME MORE PROBLEMS

1. Find $\dfrac{dy}{dx}$, from the first principle, where y is defined by

   (i) $y = x^{-3/4}$ (ii) $y = (a + bx)^{-1/3}$

2. Using the first principle, find the derivative with respect to x, of

   (i) $\tan 2x$ (ii) $\tan \sqrt{x}$ (iii) $\cos \sqrt{x}$.

3. Find the derivative with respect to x of (using the first principle)

   (i) $\cos^{-1} x^2$ (ii) $\sin(x^2 + 1)$

4. Find $\dfrac{dy}{dx}$, where y is defined by

   (i) $y = \dfrac{5x}{\sqrt{1-x^2}} + \cos^2(2x+1)$

   (ii) $y = \sqrt{\dfrac{1+e^x}{1-e^x}}$

   (iii) $y = \sin^{-1}\sqrt{\dfrac{1+x^2}{2}}$

   (iv) $y = \tan^{-1}\left[\dfrac{\sqrt{1+a^2x^2}-1}{ax}\right]$.

5. Differentiate

   (i) $\sin^{-1}\left[x\sqrt{1-x} + \sqrt{x}\sqrt{1-x^2}\right]$

   (ii) $\tan^{-1}\left[\dfrac{\sqrt{1+x^2}+\sqrt{1-x^2}}{\sqrt{1+x^2}-\sqrt{1-x^2}}\right]$ w.r.t x.

6. Find the derivatives of

   (i) $e^x \log(1 + x^2)$ (ii) $\dfrac{e^{2x} + e^{-2x}}{e^{2x} - e^{-2x}}$ (iii) $\tan^{-1} \sqrt{\dfrac{1 + \sin x}{1 - \sin x}}$ with respect to x.

7. Find $\dfrac{dy}{dx}$, when $x = \log(1 + t^2)$, $y = \tan^{-1} t$.

8. Find $\dfrac{dy}{dx}$ when $x^y = e^{x+y}$.

9. Find $\dfrac{dy}{dx}$, when

   (i) $y = \sqrt{x + \sqrt{x + \sqrt{x + \ldots \infty}}}$ (ii) $y = x^{x^{x^{x+\ldots\infty}}}$

10. Find the derivative of $\cos^{-1}\left[\dfrac{1 - x^2}{1 + x^2}\right]$, with respect to $\tan^{-1}\left[\dfrac{3x - x^3}{1 - 3x^2}\right]$.

11. Find $\dfrac{d^2y}{dx^2}$, when

    (i) $y = e^{ax} \log x$ (ii) $y = \sin^{-1} x$ (iii) $x = a\cos^2\theta$, $y = a\sin^2\theta$.

12. If $x = \sec\theta - \cos\theta$, $y = \sec^n\theta - \cos^n\theta$, then show that $(x^2 + 4)\left(\dfrac{dy}{dx}\right)^2 = n^2(y^2 + 4)$.

13. Find $\dfrac{dy}{dx}$, when

    (i) $y = \sin^{-1}(3x - 4x^3)$ (ii) $y = \sqrt{\log\left(\sin\left(\dfrac{x^2}{3} - 1\right)\right)}$.

14. Find $\dfrac{dy}{dx}$, where

    (i) $\sin y = x \sin(x + y)$ (ii) $(\tan^{-1} x)^y + y^{\cot x} = 1$.

15. Find the derivative of $\sec^{-1}\left[\dfrac{1}{2x^2 - 1}\right]$, with respect to $\sqrt{1 - x^2}$ at $x = \dfrac{1}{2}$.

ANSWERS TO SOME MORE PROBLEMS

1. (i) $-\dfrac{3}{4}x^{-7/4}$ (ii) $-\dfrac{b}{3}(a+bx)^{-4/3}$

2. (i) $2\sec^2 2x$ (ii) $\dfrac{1}{2\sqrt{x}}\sec^2\sqrt{x}$ (iii) $-\dfrac{1}{2\sqrt{2}}\sin\sqrt{x}$

3. (i) $\dfrac{-2x}{\sqrt{1-x^4}}$ (ii) $2x\cos(x^2+1)$

4. (i) $\dfrac{5}{(1-x^2)^{3/2}} - 2\sin(4x+2)$ (ii) $\dfrac{e^x}{(1+e^x)^{1/2}(1-e^x)^{3/2}}$ (iii) $\dfrac{x}{\sqrt{1-x^4}}$ (iv) $\dfrac{a}{2(1+a^2x^2)}$

5. (i) $\dfrac{1}{\sqrt{1-x^2}} + \dfrac{1}{2\sqrt{x}\sqrt{1-x}}$ (ii) $\dfrac{-x}{\sqrt{1-x^4}}$

6. (i) $e^x\left[\log(1+x^2) + \dfrac{2x}{1+x^2}\right]$ (ii) $\dfrac{-8e^{4x}}{(e^{4x}-1)^2}$,   (iii) $\dfrac{1}{2}$

7. $\dfrac{1}{2t}$ 8. $\dfrac{x-y}{x(\log x - 1)}$ 9. (i) $\dfrac{1}{2y-1}$ (ii) $\dfrac{y^2}{x(1-y\log x)}$

10. $\dfrac{2}{3}$

11. (i) $e^{ax}\left[\dfrac{2a}{x} - \dfrac{1}{x^2} + a^2\log x\right]$ (ii) $\dfrac{x}{(1-x)^{3/2}}$ (iii) 0

13. (i) $\dfrac{3}{\sqrt{1-x^2}}$ (ii) $\dfrac{x\cot\left(\dfrac{x^2}{3}+1\right)}{3\sqrt{\log\left(\dfrac{x^2}{3}-1\right)}}$

14. (i) $\dfrac{[\sin(x+y) + x\cos(x+y)]}{[\cos y - x\cos(x+y)]}$ (ii) $\dfrac{\left[y^{\cot x}\cosec^2\log y - \dfrac{(\tan^{-1}x)^y y}{(1+x^2)\tan^{-1}x}\right]}{\left[y^{\cot x - 1}\cot x + (\tan^{-1}x)^y \log\tan^{-1}(x)\right]}$

Mathsarc Education

A learning place to fulfill your dream of success!

MATHEMATICS   IIT JEE Main/Advanced

METHODS OF DIFFERENTIATION

A big goal can be achieved easily through differentiation.

CLASS WORK - UNDERSTANDING

1. $\dfrac{d^2x}{dy^2}$ equals

   (A) $\left(\dfrac{d^2y}{dx^2}\right)^{-1}$
   (B) $-\left(\dfrac{d^2y}{dx^2}\right)^{-1}\left(\dfrac{dy}{dx}\right)^{-3}$
   (C) $\left(\dfrac{d^2y}{dx^2}\right)\left(\dfrac{dy}{dx}\right)^{-2}$
   (D) $-\left(\dfrac{d^2y}{dx^2}\right)\left(\dfrac{dy}{dx}\right)^{-3}$

2. Let f(x) and g(x) are differentiable functions such that $\dfrac{f(x)}{g(x)}=7$. If $\dfrac{f'(x)}{g'(x)}=\alpha$ and $\left(\dfrac{f(x)}{g(x)}\right)'=\beta$ (f'(x) represents the derivative of f(x) w.r.t. x), then $\dfrac{\alpha-\beta}{\alpha+\beta}=$ _____

   (A) 0   (B) 1   (C) 7   (D) None of these

3. Let $f(x) = (x^2 - 3x + 2)\,|x^3 - 6x^2 + 11x - 6| + \left|\sin\left(x+\dfrac{\pi}{4}\right)\right|$. Number of points at which the function f(x) is non-differentiable in [0, 2π], is

   (A) 5   (B) 4   (C) 3   (D) 2

4. If $f(x) = \begin{cases} x\sin\left(\dfrac{1}{x}\right) & \text{for } x \neq 0 \\ 0 & \text{for } x = 0 \end{cases}$, then

   (A) Both f'(0⁺) and f'(0⁻) Do not Exist
   (B) f'(0⁺) exist but f'(0⁻) does not
   (C) f'(0⁺) = f'(0⁻)
   (D) None of these

5. If $f(x) = \cos x - \int_0^x (x-t)f(t)dt$, then $f''(x) + f(x)$ equals

   (A) $-\cos x$ (B) 0 (C) $\int_0^x (x-t)f(t)dt$ (D) $-\int_0^{-x}(x-t)f(t)dt$

6. If $f(x)$ is differentiable and $\int_0^{t^2} x f(x)dx = \frac{2}{5}t^5$, then $f\left(\frac{4}{25}\right)$ equals

   (A) $2/5$ (B) $-5/2$ (C) 1 (D) $5/2$

7. If $f(x) = \int_{-1}^{1} \frac{\sin x}{1+t^2}dt$, then $f'\left(\frac{\pi}{3}\right)$ is

   (A) Non - existent (B) $\frac{\pi}{4}$ (C) $\frac{\pi\sqrt{3}}{4}$ (D) none of these

8. If $f(x) = \frac{1}{x^2}\int_4^x (4t^2 - 2f'(t))dt$ then $f'(4)$ is equal to

   (A) 16 (B) $\frac{32}{9}$ (C) $\frac{32}{3}$ (D) none of these

9. Let $f(x)$ be a differentiable function such that $f'(x) + f(x) = 4xe^{-x}\cdot \sin 2x$ and $f(0) = 0$. Then the value of $\lim_{n\to\infty}\sum_{k=1}^{n}f(k\pi)$ is/are

   (A) $-\dfrac{2\pi e^{\pi}}{(e^{\pi}-1)^2}$ (B) $\dfrac{2\pi e^{\pi}}{(e^{\pi}-1)^2}$ (C) $-\dfrac{2\pi e^{\pi}}{(e^{\pi}+1)^2}$ (D) $\dfrac{2\pi e^{\pi}}{(e^{\pi}+1)^2}$

10. If $x = f(t)$ and $y = g(t)$ are differentiable functions of t then $\dfrac{d^2y}{dx^2}$ is

    (A) $\dfrac{f'(t)g''(t) - g'(t)f''(t)}{(f'(t))^3}$ (B) $\dfrac{f'(t)g''(t) - g'(t)f''(t)}{(f'(t))^2}$

    (C) $\dfrac{g'(t)f''(t) - f'(t)g''(t)}{(f'(t))^3}$ (D) $\dfrac{g'(t)f''(t) + f'(t)g''(t)}{(f'(t))^3}$

11. Let f: R→R be a continuous & differentiable function given by $f(x) = x + \int_0^1 (xy + x^2) f(y) dy$. Then

(A) $\int_0^1 f(x) dx = \dfrac{26}{23}$ (B) $\int_0^1 f(x) dx = \dfrac{25}{13}$ (C) $\int_0^1 x f(x) dx = \dfrac{13}{25}$ (D) $\int_0^1 x f(x) dx = \dfrac{25}{23}$

12. If $f(x) = \int_0^{g(x)} \dfrac{dt}{\sqrt{1+t^3}}$, $g(x) = \int_0^{\cos x} (1 + \sin t^2) dt$ then the value of $f'\left(\dfrac{\pi}{2}\right) =$

(A) 1   (B) −1   (C) 0   (D) ½

13. If g is the inverse of f & $f'(x) = \dfrac{1}{1+x^5}$ then $g'(x)$ equals ___

(A) $1 + [g(x)]^5$ (B) $\dfrac{1}{1+[g(x)]^5}$ (C) $-\dfrac{1}{1+[g(x)]^5}$ (D) None of these

14. If $y = f\left(\dfrac{3x+4}{5x+6}\right)$ & $f'(x) = \tan(x^2)$ then $\dfrac{dy}{dx} =$

(A) $\tan(x^3)$ (B) $-2\tan\left[\dfrac{3x+4}{5x+6}\right]^2 \times \dfrac{1}{(5x+6)^2}$

(C) $f\left(\dfrac{3\tan(x^2)+4}{5\tan(x^2)+6}\right) \cdot \tan(x^2)$ (D) None of these

15. Let $f(x) = \begin{vmatrix} x^3 & \sin x & \cos x \\ 6 & -1 & 0 \\ p & p^2 & p^3 \end{vmatrix}$ where p is a constant. Then $\dfrac{d^3[f(x)]}{dx^3}\bigg|_{x=0}$ is:

(A) $6p^3$ (B) $p + p^2$ (C) $p + p^3$ (D) independent of p

16. If $f(x) = (2x - 3\pi)^5 + \dfrac{4x}{3} + \cos x$ and g is the inverse function of f, then $g'(2\pi)$ is equal to:

(A) 7/3   (B) 3/7   (C) $\dfrac{30\pi^4 + 4}{3}$ (D) $\dfrac{3}{30\pi^4 + 4}$

17. If $y = \dfrac{\cos 6x + 6\cos 4x + 15\cos 2x + 10}{\cos 5x + 5\cos 3x + 10\cos x}$, then $\dfrac{dy}{dx} = $ ....

(A) $2\sin x + \cos x$ (B) $-2\sin x$ (C) $\cos 2x$ (D) $\sin 2x$

18. $\lim\limits_{x \to 0^+}\left(x^{x^x} - x^x\right)$ is equal to:

(A) 0   (B) 1   (C) -1 (D) D.N.E.

19. If $x_1, x_1, x_2, x_3, \ldots, x_{n-1}$ be n zero's of the polynomial $P(x) = x^n + \alpha x + \beta$, where $x_i \neq x_j \; \forall \; i, j \in \{1, 2, 3, \ldots, n-1\}$. The value of $Q(x) = (x_1 - x_2)(x_1 - x_3)(x_1 - x_4)\ldots(x_1 - x_{n-1})$, is:

(A) $n(n-1)x_1^{n-2}$ (B) $^nC_2 x_1^{n-2}$ (C) $n x_1^{n-1} + \alpha$ (D) Zero

20. If $\sin x = \dfrac{2t}{1+t^2}$ & $\cot y = \dfrac{1-t^2}{2t}$, then the value of $\dfrac{d^2x}{d^2y}$, is equal to:

(A) 0   (B) 1   (C) -1 (D) $1/2$

21. The value of $\lim\limits_{x \to 0^+}\left(x^x + (\tan x)^{\cosec x} + (\cosec x)^{\tan x}\right)$ is equal to:

(A) 1   (B) 2   (C) $2 + \dfrac{1}{e}$ (D) $1 + \dfrac{1}{e}$

22. If a differentiable function $f(x) = e^x + 2x$ is given, then $\dfrac{d}{dx}\left(f^{-1}(x)\right)$ at $x = f(\ln 3)$ is equal to:

(A) $1/5$ (B) $3/7$ (C) $7/3$ (D) 5

23. For the curve $32x^3 y^2 = (x + y)^5$, the value of $\dfrac{d^2y}{dx^2}$ at $P(1, 1)$ is equal to:

(A) 0   (B) 1   (C) -1 (D) $1/2$

24. Let $f: (-2, 2) \to R$ be a differentiable function such that $f(0) = -1$ and $f'(0) = 1$. If $g(x) = \left(f(2f(x) + 2)\right)^2$ then $g'(0)$ is equal to:

(A) -4 (B) 0   (C) -2 (D) 4

25. Let $f(x) = \log_3\left(\dfrac{1-x}{1+x}\right) + \log_3\left(x + \sqrt{x^2+1}\right)$ then:

(A) The graph, $y = f(x)$ symmetric about y − axis   (B) $f(0) = 1$
(C) $f'(0) = 0$ (D) $f''(0) = 0$

26. Let A, B, P be the points the curve y = lnx with their x co-ordinate as 1, 2 and t respectively then the value of $\lim_{t\to\infty} \cos\angle BAP$ is:

(A) $\sqrt{1+\ln^2 2}$ (B) ln 2 (C) $\dfrac{1}{\sqrt{1+\ln^2 2}}$ (D) $\dfrac{1}{1+\ln 2}$

27. If $8f(x)+6f\left(\dfrac{1}{x}\right)=x+5$ and $y = x^2 f(x)$, then $\left.\dfrac{dy}{dx}\right|_{x=-1}$ is equal to:

(A) 0 (B) $\dfrac{1}{14}$ (C) $-\dfrac{1}{14}$ (D) none of these

28. If $x^p x^q = (x+y)^{p+q}$ then $\dfrac{dy}{dx}$ is

(A) independent of p

(B) independent of q

(C) dependent on both p & q (D) $\dfrac{y}{x}$

29. Let f: R→R, g: R→R and h: R→R be differentiable function such that $f(x) = x^3 + 2x + 1$, g (f(x)) = x and h (g (g(x))) = x ∀ x ∈ R. Then

(A) g'(1) = 1/2   (B) h'(0) = 10

(C) If x_0 ∈ R, $x_0^3 + 2x_0 - 2 = 0$ then $h(x_0) = 34$ (D) h(g(2)) = 12

30. Let a, b ∈ R and f: R→ R be defined by $f(x)=a\cos(|x^3-x|)+b|x|\sin(|x^3+x|)$. Then f is___

(A) Differentiable at x = 0 if a = 0 and b = 1

(B) Differentiable at x = 1 if a = 1 and b = 0

(C) **NOT** differentiable at x = 0 if a = 1 and b = 0

(D) **NOT** differentiable at x = 1 if a = 1 and b = 1

31. If $\Delta(x)=\begin{vmatrix} x^2+4x-3 & 2x+4 & 13 \\ 2x^2+5x-9 & 4x+5 & 26 \\ 8x^2-6x+1 & 16x-6 & 104 \end{vmatrix}=ax^3+bx^2+cx+d$, then

(A) a = 3 (B) b = 0 (C) c = 0 (D) none of these

32. Let $f:\left[-\frac{1}{2},2\right]\to R$ & $g:\left[-\frac{1}{2},2\right]\to R$ be function defined by $f(x)=[x^2-3]$ & $g(x)=|x|f(x)+|4x-7|f(x)$, where $[y]$ = G.I.F for $y \in R$. then

(A) f is discontinuous exactly at three points in $\left[-\frac{1}{2},2\right]$

(B) f is discontinuous exactly at four points in $\left[-\frac{1}{2},2\right]$

(C) g is NOT differentiable exactly at four points in $\left(-\frac{1}{2},2\right)$

(D) g is NOT differentiable exactly at five points in $\left(-\frac{1}{2},2\right)$

33. Given $\dfrac{\int_{f(y)}^{f(x)} e^t dt}{\int_y^x \frac{1}{t} dt}=1 \ \forall\ x,y \in \left(\frac{1}{e^2},\infty\right)$ Where $f(x)$ is continuous & differentiable function s.t. $f\left(\frac{1}{e}\right)=0$. If $g(x)=\begin{cases} e^x, & x \geq k \\ e^{x^2}, & 0<x<k\end{cases}$, then

(A) $f(g(x))$ is continuous for $k=1$
(B) $f(g(x))$ is differentiable for $k=1$
(C) $f(g(x))$ is non-differentiable for $k=1$
(D) $f(g(x))$ is continuous for $k=2$

34. The function $f(x)=\begin{cases} |x-3|, & x \geq 1 \\ \frac{x^2}{4}-\frac{3x}{2}+\frac{13}{4}, & x<1 \end{cases}$ is

(A) Continuous at x = 1
(B) differentiable at x = 1
(C) Continuous at x = 3
(D) differentiable at x = 3

35. Let 'f' be a differentiable function satisfying $f(x+y)=f(x)+f(y)+(e^x-1)(e^y-1)\ \forall\ x,y \in R$ and $f'(0)=2$. Identify the correct statement(s):

(A) $\lim\limits_{x\to 0}\dfrac{f(f(x))}{f(x)-x}=4$
(B) $\lim\limits_{x\to 0}(f(x)+\cos x)^{1/(e^x-1)}=e^2$

(C) Number of roots of equation f(x) = 0 are 2   (D) Range of f(x) is $(-\infty, \infty)$

36. If $f(x) = x^n$ then find the value of $f(1) + \dfrac{f^1(1)}{1!} + \dfrac{f^2(1)}{2!} + \ldots + \dfrac{f^n(1)}{n!}$ where $f^r(x)$ denotes rth derivative of f(x) w.r.t. x.

37. Let f(x) be a differentiable function such that $f(x) = 1 + \dfrac{x^3}{3} + \int_0^x e^{-t} f(x-t) dt$, if $\int_0^1 f(x) dx = p$ then the value of 5p is _____

38. Let f(x) be a differentiable function in $[-1, \infty)$ and $f(0) = 1$ such that $\lim\limits_{t \to x+1} \dfrac{t^2 f(x+1) - (x+1)^2 f(t)}{f(t) - f(x+1)} = 1$. Find the value of $\lim\limits_{x \to 1} \dfrac{\ln(f(x)) - \ln 2}{x - 1}$.

39. If $2x = (y^{1/3} + y^{-1/3})$, then find the value of $\dfrac{(x^2-1)}{y} \cdot \dfrac{d^2 y}{dx^2} + \dfrac{x}{y} \cdot \dfrac{dy}{dx}$

40. Evaluate: $\lim\limits_{x \to 0} \left(\dfrac{1}{x^5} \int_0^x e^{-t^2} dt - \dfrac{1}{x^4} + \dfrac{1}{3x^2} \right)$

41. If P_n is the sum of GP upto n terms. Show that $(1-r)\dfrac{dP_n}{dr} = nP_{n-1} - (n-1)P_n$.

42. Let $f(x) = x + \dfrac{1}{2x + \dfrac{1}{2x + \dfrac{1}{2x + \ldots \infty}}}$. Compute the value of $f(5)f'(5)$ ____

43. Differentiate: $\dfrac{\sqrt{1+x^2} + \sqrt{1-x^2}}{\sqrt{1+x^2} - \sqrt{1-x^2}}$ w.r.t. $\sqrt{1-x^4}$.

44. (a) Let $f(x) = x^2 - 4x - 3$, $x > 2$ and let g be the inverse of f. Find the value of g' where f(x) = 2.

    (b) Let f, g and h are differentiable functions. If $f(0) = 1$; $g(0) = 2$; $h(0) = 3$; and the derivatives of their pairwise product at x = 0 are $(fg)'(0) = 6$; $(gh)'(0) = 4$ and $(hf)'(0) = 5$ then compute the value of $(fgh)'(0)$.

45. If $f : R \to R$ is a function such that $f(x) = x^3 + x^2 f'(1) + x f''(2) + f'''(3)$ for all $x \in R$, then prove that $f(2) = f(1) - f(0)$.

46. If the function $f(x) = x^3 + e^{x/2}$ and $g(x) = f^{-1}(x)$, then the value of $g'(1)$ is ____

47. Let $y'(x) + y(x)g'(x) = g(x)g'(x)$, $y(0) = 0$, $x \in R$, where $f'(x)$ denotes $\dfrac{df(x)}{dx}$ and $g(x)$ is a given non-constant differentiable function on R with $g(0) = g(2) = 0$.

    Then the value of $y(2)$ is ____

48. Consider $f(x) = (x + 1)(x + 2)(x + 3)\ldots(x + n)$. Find $f'(0)$.

49. If $y = \dfrac{x^4 - x^2 + 1}{x^2 + \sqrt{3}x + 1}$ and $\dfrac{dy}{dx} = ax + b$ then find the value of $a + b$ ____

50. If $y = (\ln x)^{(\ln x)^{(\ln x)^{\cdots \infty}}}$. Find $\dfrac{dy}{dx}$

51. A function f satisfies the relation $f(x) = f''(x) + f'''(x) + f''''(x) + \ldots \infty$ where $f(x)$ is a differentiable function indefinitely. If $f(1) = 5$, then the value of $f'(1) + f''(1)$ is equal to

52. If $\int_0^x f(t)dt = x^2 + \int_x^1 t^2 f(t)dt$, then $f'\left(\dfrac{1}{2}\right) =$ _____

Answer Key - Differentiation

1. D	2. B	3. C	4. A
5. A	6. A	7. B	8. B
9. A	10. A	11. D	12. B
13. A	14. B	15. D	16. B
17. B	18. C	19. B	20. A
21. B	22. A	23. A	24. A
25. D	26. C	27. C	28. D
29. A, B, C	30. A, B	31. B, C	32. B, C
33. (A,B,C)	34. A, B, C	35. A, B, D	36. 2^n
37. 8	38. 1	39. 9	40. 1/10
42. 5	43. $\dfrac{1+\sqrt{1+x^4}}{x^6}$	44. (a) 1/6 (b) 16	46. 2
47. 0	48. $n!\left(1+\dfrac{1}{2}+\dfrac{1}{3}+\ldots+\dfrac{1}{n}\right)$	49. $2-\sqrt{3}$	
50. $\dfrac{dy}{dx}=\dfrac{y^2}{x(1-y\ln x)}$	51. 5	52. 24/25	

Mathsarc Education

A learning place to fulfill your dream of success!

MATHEMATICS   IIT JEE Main/Advanced

MATRIX & DETERMINANT

Multi-dimensional analysis in one go. Helping the world in analytics.

SINGLE OPTION CORRECT

1. If $x^a y^b = e^m$, $x^c y^d = e^n$, $\Delta_1 = \begin{vmatrix} m & b \\ n & d \end{vmatrix}$, $\Delta_2 = \begin{vmatrix} a & m \\ c & n \end{vmatrix}$, $\Delta_3 = \begin{vmatrix} a & b \\ c & d \end{vmatrix}$, then the values of x and y are respectively.

   (A) $\dfrac{\Delta_1}{\Delta_3}$ and $\dfrac{\Delta_2}{\Delta_3}$

   (B) $\dfrac{\Delta_2}{\Delta_1}$ and $\dfrac{\Delta_3}{\Delta_1}$

   (C) $\log\left(\dfrac{\Delta_1}{\Delta_3}\right)$ and $\log\left(\dfrac{\Delta_2}{\Delta_3}\right)$

   (D) $e^{\frac{\Delta_1}{\Delta_3}}$ and $e^{\frac{\Delta_2}{\Delta_3}}$

2. The system of equations
$$\left.\begin{array}{l} \alpha x + y + z = \alpha - 1 \\ x + \alpha y + z = \alpha - 1 \\ x + y + \alpha z = \alpha - 1 \end{array}\right\}$$, has no solution, if α is

   (A) 1   (B) not -2   (C) either -2 or 1   (D) -2

3. Let $\omega = -\dfrac{1}{2} + i\dfrac{\sqrt{3}}{2}$, then the value of determinant $\begin{vmatrix} 1+\omega & \omega^2 & -\omega \\ 1+\omega^2 & \omega & -\omega^2 \\ \omega^2+\omega & \omega & -\omega^2 \end{vmatrix}$ is

   (A) -3   (B) $-3\omega^2$ (C) $3\omega^2$ (D) 3

4. If $A = \begin{bmatrix} 1 & -1 & 1 \\ 0 & 2 & -3 \\ 2 & 1 & 0 \end{bmatrix}$ and B = adj(A) and C = 5A, then $\dfrac{|\text{adj} B|}{|C|}$ is equal to

   (A) 5   (B) 25   (C) -1   (D) 1

5. If P is a 3×3 matrix such that $P^T = 2P+I$, where P^T is the transpose of P and I is the 3×3 identity matrix, then there exist a column matrix $X = \begin{bmatrix} x \\ y \\ z \end{bmatrix} \neq \begin{bmatrix} 0 \\ 0 \\ 0 \end{bmatrix}$ such that

   (A) $PX = \begin{bmatrix} 0 \\ 0 \\ 0 \end{bmatrix}$
   (B) PX = X
   (C) PX = 2X
   (D) PX = - X

6. The number of different non-singular matrices of the type $A = \begin{bmatrix} 1 & a & c \\ 1 & 1 & b \\ 0 & -w & w \end{bmatrix}$ where $w = e^{i\theta}$ and $a, b, c \in \{z : z^4 - 1 = 0\}$ are

   (A) 44
   (B) 48
   (C) 56
   (D) 55

7. If adj(B) = A, |P| = |Q| = 1, then adj($Q^{-1} B P^{-1}$) is:

   (A) PQ
   (B) QAP
   (C) PAQ
   (D) $PA^{-1}Q$

8. Let $P = [a_{ij}]$ be a 3×3 matrix and let $Q = [b_{ij}]$ where $b_{ij} = 2^{i+j} a_{ij}$ for $1 \le i, j \le 3$. If the determinant of P is 2, then the determinant of matrix Q is

   (A) 2^{10}
   (B) 2^{11}
   (C) 2^{12}
   (D) 2^{13}

9. How many 3×3 matrices M with entries from {0, 1, 2} are there, for which the sum of the diagonal entries of $M^T M$ is 5?

   (A) 126
   (B) 198
   (C) 162
   (D) 135

10. Let matrix $A = \begin{bmatrix} x & y & -z \\ 1 & 2 & 3 \\ 1 & 1 & 2 \end{bmatrix}$, where $x, y, z \in N$. If $|adj(adj(adj(adjA)))| = 4^8 \cdot 5^{16}$, then the number of such matrices are ?

    (A) 28
    (B) 45
    (C) 36
    (D) 55

11. If $Z = \begin{vmatrix} 3+2i & 1 & i \\ 2 & 3-2i & 1+i \\ 1-i & -i & 3 \end{vmatrix}$ and $|z + \bar{z}| = k|z|$, then 3k is

    (A) 0
    (B) 6
    (C) 3
    (D) 9

12. If $\begin{vmatrix} x & 3 & x \\ x^2 & x & 6 \\ x & x & 6 \end{vmatrix} = Ax^4 + Bx^3 + Cx^2 + Dx + E$, then the value of 5A + 4B + 3C + 2D + E is _____

   (A) 0 (B) 15 (C) -16 (D) -17

13. If $A = \begin{bmatrix} 2 & -2 & -4 \\ -1 & 3 & 4 \\ 1 & -2 & -3 \end{bmatrix}$ and $B = \begin{bmatrix} -4 & -3 & -3 \\ 1 & 0 & 1 \\ 4 & 4 & 3 \end{bmatrix}$, then the value of

   $|A + A^2B^2 + A^3 + A^4B^4 + ..100 \text{ terms}|$ is equal to

   (A) 1000 (B) -800 (C) 0 (D) -8000

14. If $P = \begin{bmatrix} \frac{\sqrt{3}}{2} & \frac{1}{2} \\ -\frac{1}{2} & \frac{\sqrt{3}}{2} \end{bmatrix}$, $A = \begin{bmatrix} 1 & 1 \\ 0 & 1 \end{bmatrix}$ and $Q = PAP^T$ and $X = P^T Q^{2005} P$, then X is equal to

   (A) $\begin{bmatrix} 1 & 2005 \\ 0 & 1 \end{bmatrix}$ (B) $\begin{bmatrix} 4+2005\sqrt{3} & 6015 \\ 2005 & 4-2005\sqrt{3} \end{bmatrix}$

   (C) $\frac{1}{4}\begin{bmatrix} 2+\sqrt{3} & 1 \\ -1 & 2-\sqrt{3} \end{bmatrix}$ (D) $\frac{1}{4}\begin{bmatrix} 2005 & 2-\sqrt{3} \\ 2+\sqrt{3} & 2005 \end{bmatrix}$

15. Let A be a 2×3 matrix whereas B be a 3×2 matrix. If det (AB) = 4, then value of det (BA), is?

   (A) 1 (B) -1 (C) 0 (D) 5

16. Evaluate: $\||A^T|A\|A\|$ where $A = \begin{bmatrix} 1 & 7 & -1 \\ 2 & 3 & 2 \\ 3 & -1 & 4 \end{bmatrix}$.

   (A) 11^{13} (B) 13^{11} (C) 11^9 (D) None of these

17. A and B are matrices of order 3×2 and 2×3 respectively. If $AB = \begin{bmatrix} 8 & 2 & -2 \\ 2 & 5 & 4 \\ -2 & 4 & 5 \end{bmatrix}$ then

   (A) $BA = \begin{bmatrix} 0 & 9 \\ 9 & 0 \end{bmatrix}$ (B) $BA = \begin{bmatrix} 9 & 0 \\ 0 & 9 \end{bmatrix}$ (C) ABAB = 9AB (D) ABAB = AB

MULTI OPTION(S) CORRECT

18. If A and B are 3×3 matrices and $|A| \neq 0$, then which of the following are true?

    (A) $|AB| = 0 \to |B| = 0$
    (B) $|AB| = 0 \to B = O$
    (C) $|A^{-1}| = |A|^{-1}$
    (D) $|A + A| = 2|A|$

19. If $A = \begin{bmatrix} a & b \\ c & d \end{bmatrix}$ (where $bc \neq 0$) satisfies the equations $x^2 + k = 0$, then

    (A) $a + d = 0$
    (B) $k = -|A|$
    (C) $k = |A|$
    (D) None of these

20. Let $A = \begin{bmatrix} 1 & 2 & 2 \\ 2 & 1 & 2 \\ 2 & 2 & 1 \end{bmatrix}$, then

    (A) $A^2 - 4A - 5I_3 = O$
    (B) $A^{-1} = \dfrac{1}{5}(A - 4I_3)$
    (C) A^3 is not invertible
    (D) A^2 is invertible

21. Let $P = \begin{bmatrix} 3 & -1 & -2 \\ 2 & 0 & \alpha \\ 3 & -5 & 0 \end{bmatrix}$, where $\alpha \in R$, Suppose $Q = [q_{ij}]$ is a matrix such that $PQ = kI$, where $k \in R$, $k \neq 0$ and I is the identity matrix of order 3. If $q_{23} = -\dfrac{k}{8}$ and $\det(Q) = \dfrac{k^2}{2}$, then –

    (A) $\alpha = 0, k = 8$
    (B) $4\alpha - k + 8 = 0$
    (C) $\det(P \, adj(Q)) = 2^9$
    (D) $\det(Q \, adj(P)) = 2^{13}$

22. Let $0 < \theta < \dfrac{\pi}{2}$ and $\Delta(x,\theta) = \begin{vmatrix} x & \tan\theta & \cot\theta \\ -\tan\theta & -x & 1 \\ \cot\theta & 1 & x \end{vmatrix}$ then

    (A) $\Delta(0, \theta) = 0$
    (B) $\Delta\left(x, \dfrac{\pi}{4}\right) = x^2 + 1$
    (C) $\text{Min}\Delta(1, \theta) = 0$ when $\theta \in \left(0, \dfrac{\pi}{2}\right)$
    (D) $\Delta(x, \theta)$ is independent on x.

23. Select the correct statements for the matrix $A = \begin{bmatrix} -1 & 3 & -5 \\ 2 & 1 & 7 \\ 0 & 6 & 1 \end{bmatrix}$

(A) $|adj(A)| = 625$

(B) $adj\left(\frac{1}{5}A\right) = \frac{1}{25}adj(A)$

(C) $adj(A^{-1}) = -\frac{1}{25}A$

(D) $\begin{bmatrix} -1 & 2 & 0 \\ 3 & 1 & 6 \\ -5 & 7 & 1 \end{bmatrix}^{-1} = (A^{-1})^T$

24. Let A and B be two square matrix of order 3, then which of the following statement(s) is/are correct?

(A) ABA^T is a symmetric matrix

(B) $AB - BA$ is a skew symmetric matrix

(C) If $B = |A|A^{-1}$, $|A| \neq 0$, then $adj(A^T) - B$ is a skew symmetric matrix

(D) If $B + A^T = O$ and A is a skew symmetric matrix, then B^{15} is also skew symmetric matrix

25. Which of the following is (are) NOT the square of a 3×3 matrix with real entries?

(A) $\begin{bmatrix} 1 & 0 & 0 \\ 0 & 1 & 0 \\ 0 & 0 & 1 \end{bmatrix}$
(B) $\begin{bmatrix} 1 & 0 & 0 \\ 0 & 1 & 0 \\ 0 & 0 & -1 \end{bmatrix}$
(C) $\begin{bmatrix} 1 & 0 & 0 \\ 0 & -1 & 0 \\ 0 & 0 & -1 \end{bmatrix}$
(D) $\begin{bmatrix} -1 & 0 & 0 \\ 0 & -1 & 0 \\ 0 & 0 & -1 \end{bmatrix}$

26. If $A^{10} = \begin{bmatrix} a & b \\ c & d \end{bmatrix}$, where $A = \begin{bmatrix} 2 & 1 \\ 0 & 3 \end{bmatrix}$ then

(A) Number of Natural factors of a are 11

(B) b is an integer

(C) Number of Natural factors of $a+b+c+d$ are 22

(D) $a + d$ is a multiple of 13

27. Let $A = \begin{bmatrix} 1 & 0 \\ 1 & 1 \end{bmatrix}$ then _____ {Where $A^{-n} = (A^{-1})^n$}

(A) $A^{-n} = \begin{bmatrix} 1 & 0 \\ -n & 1 \end{bmatrix} \forall n \in N$
(B) $\lim_{n\to\infty} \frac{A^{-n}}{n} = \begin{bmatrix} 0 & 0 \\ -1 & 0 \end{bmatrix}$
(C) $\lim_{n\to\infty} \frac{A^{-n}}{n^2} = \begin{bmatrix} 0 & 0 \\ 0 & 0 \end{bmatrix}$
(D) $\det(A^{-n}) = 1$

INTEGER OPTION TYPE (0 - 9)

28. Consider, $f(x) = \dfrac{-x^2}{(x^2-9)(x-7)^2(x-9)(x-3)}$ where a_i are the integral values of x for which

    $f(x) \geq 0$ and $a_i < a_{i+1}\ \forall i = 1,2,\ldots,8$. If $A = \begin{bmatrix} a_1 & a_2 & a_3 \\ a_4 & a_5 & a_6 \\ a_7 & a_8 & a_9 \end{bmatrix}$, and $B^3 - pB^2 + qB - rI = O$

    Where $B = adj.A$, then find the value of $(2r + p)$.

29. Let $A = \begin{bmatrix} 3 & 2 & 1 \\ 6x^2 & 2x^3 & x^4 \\ 1 & a & a^2 \end{bmatrix}$ then $\dfrac{d^2}{dx^2}\left(|A(adjA)|^{\frac{1}{3}}\right)$ at x = a is......

30. Let $A = [a_{ij}](1 \leq i,j \leq 3)$ be a 3×3 matrix and $B = [b_{ij}](1 \leq i,j \leq 3)$ be a 3×3 matrix such that $b_{ij} = \sum_{k=1}^{3} a_{ik}a_{jk}$. If $\det(A) = 2$, then the value of $\det(B)$ is _____

31. Let (x, y, z) be points with integer co-ordinates satisfying the system of homogeneous equation x + y + z = 0, x + 2y + 3z = 0 and 2x + 3y + 4z = 0, then the number of such points for which $x^2 + y^2 + z^2 \leq 12$.

32. A be a square matrix of order 2 with $|A| \neq 0$ such that $|A + |A|\ adj(A)| = 0$, then the value of $|A - |A|\ adj(A)|$ is?

33. Consider the matrices $A = \begin{bmatrix} 2 & 1 \\ 4 & 1 \end{bmatrix}$, $B = \begin{bmatrix} 3 & 4 \\ 2 & 3 \end{bmatrix}$ and $C = \begin{bmatrix} 3 & -4 \\ -2 & 3 \end{bmatrix}$, then the value of

    $tr(A) + tr\left(\dfrac{ABC}{2}\right) + tr\left(\dfrac{A(BC)^2}{4}\right) + \ldots \infty$ is ? {Where tr(A) = sum of diagonal elements}

34. If A is a square matrix of order 3, then find $\left|(A - A^T)^{105}\right|$.

35. If $3A = \begin{bmatrix} 1 & 2 & 2 \\ 2 & 1 & -2 \\ x & 2 & y \end{bmatrix}$ and $AA^T = I_3$ then $|x + y|$ is equal to _____

36. For a real number α, if the system $\begin{bmatrix} 1 & \alpha & \alpha^2 \\ \alpha & 1 & \alpha \\ \alpha^2 & \alpha & 1 \end{bmatrix}\begin{bmatrix} x \\ y \\ z \end{bmatrix} = \begin{bmatrix} 1 \\ -1 \\ 1 \end{bmatrix}$ of linear equations, has infinitely many solutions, then $1 + \alpha + \alpha^2 =$

37. Let the matrix $A = \begin{bmatrix} 1 & 2 & 2 \\ 2 & 1 & 2 \\ 2 & 2 & 1 \end{bmatrix}$ be a zero divisor of the polynomial $f(x) = x^2 - 4x - 5$. If the sum of all the elements in the matrix A^3 is $3(P)^3$, find the value of P.

38. Total number of distinct $x \in R$ for which $\begin{vmatrix} x & x^2 & 1+x^3 \\ 2x & 4x^2 & 1+8x^3 \\ 3x & 9x^2 & 1+27x^3 \end{vmatrix} = 10$

39. Let $z = \dfrac{-1+i\sqrt{3}}{2}$, where $i = \sqrt{-1}$ and $r, s \in \{1, 2, 3\}$. Let $P = \begin{bmatrix} (-z)^r & z^{2s} \\ z^{2s} & z^r \end{bmatrix}$ and I be the identity matrix of order 2. Then the total number of ordered pairs (r, s) for which $P^2 = -I$ is ____

40. Match The column

	COLUMN - I		COLUMN - II
(A)	$(adj\ A)^{-1}$	(P)	$k^{n-1} (adj\ A)$
(B)	$adj\ (A^{-1})$	(Q)	$\dfrac{A}{|A|}$
(C)	$adj(kA)$	(R)	$|A|^{n-2} A$
(D)	$adj(adj\ A)$	(S)	$\dfrac{adj(adj\ A)}{|A|^2}$

SUBJECTIVE PROBLEMS

1. Find α, β, γ if $\begin{bmatrix} 0 & 2\beta & \gamma \\ \alpha & \beta & -\gamma \\ \alpha & -\beta & \gamma \end{bmatrix}$ is orthogonal matrix.

2. Let a system of equation is given as

   $\left.\begin{array}{l} 2x + py + 6z = 8 \\ x + 2y + qz = 5 \\ x + y + 3z = 4 \end{array}\right\}$ Then find the value of p & q if the system of equations has

   (i) No Solution   (ii) Unique Solution   (iii) Infinite Many Solution

3. Let there are 40 different elements, then find

   (i) Total number of possible orders of matrices which can be formed using these elements.

   (ii) Total number of matrices which can be formed using all of these elements.

4. Find the cardinality of set A (i.e. n(A)), where A = {x: x = $[a_{ij}]_{2\times 2}$, $a_{ij} \in \{0, 1, 2\}$ & det(x) = 0}.

   Given, If $A = \begin{bmatrix} a & b \\ c & d \end{bmatrix}$, then det(A) = ad – bc.

5. Find the number of 2 × 2 matrix satisfying the following conditions

   (i) a_{ij} is 1 or –1   (ii) $a_{11}^2 + a_{12}^2 = a_{21}^2 + a_{22}^2 = 2$ (iii) $a_{11} a_{21} + a_{12} a_{22} = 0$

6. Find the matrix A satisfying the matrix equation, $\begin{bmatrix} 2 & 1 \\ 3 & 2 \end{bmatrix} \cdot A \cdot \begin{bmatrix} 3 & 2 \\ 5 & -3 \end{bmatrix} = \begin{bmatrix} 2 & 4 \\ 3 & -1 \end{bmatrix}$.

7. Find the product of two matrices A & B, where $A = \begin{bmatrix} -5 & 1 & 3 \\ 7 & 1 & -5 \\ 1 & -1 & 1 \end{bmatrix}$ & $B = \begin{bmatrix} 1 & 1 & 2 \\ 3 & 2 & 1 \\ 2 & 1 & 3 \end{bmatrix}$ & use it

   to solve the given system of linear equations x + y + 2z = 1, 3x + 2y + z = 7 & 2x + y + 3z = 2.

8. Determine the values of a and b for which the system $\begin{bmatrix} 3 & -2 & 1 \\ 5 & -8 & 9 \\ 2 & 1 & a \end{bmatrix} \begin{bmatrix} x \\ y \\ z \end{bmatrix} = \begin{bmatrix} b \\ 3 \\ -1 \end{bmatrix}$

   (i) has a unique solution (ii) has no solution and (iii) has infinitely many solutions

9. If matrix $A = \begin{bmatrix} a & b & c \\ b & c & a \\ c & a & b \end{bmatrix}$ where a, b, c are real numbers such that abc = 1 and $A^T A = I$, then find the value of $a^3 + b^3 + c^3$.

10. Let $A = \begin{bmatrix} 1 & -1 & 1 \\ 2 & 1 & -3 \\ 1 & 1 & 1 \end{bmatrix}$ & $10B = \begin{bmatrix} 4 & 2 & 2 \\ -5 & 0 & \alpha \\ 1 & -2 & 3 \end{bmatrix}$. If B is the inverse of A, then find the value of α

11. Find the adjoint of the matrix $A = \begin{bmatrix} 1 & 2 & 3 \\ 1 & 3 & 5 \\ 1 & 5 & 12 \end{bmatrix}$ and hence, evaluate A^{-1}.

12. For the matrix $A = \begin{bmatrix} 1 & -1 & 1 \\ 2 & 3 & 0 \\ 18 & 2 & 10 \end{bmatrix}$, show that A (adj A) = O.

13. Investigate for what values of λ, μ the simultaneous equations.

    $x + y + z = 6$, $x + 2y + 3z = 10$, $x + 2y + \lambda z = \mu$ have

    (i) no solution   (ii) a unique solution   (iii) an infinite number of solutions.

14. Let λ and α be real. Find the set of all values of λ for which the system of linear equations

    $\lambda x + \sin\alpha . y + \cos\alpha . z = 0$, $x + \cos\alpha . y + \sin\alpha . z = 0$, $-x + \sin\alpha . y - \cos\alpha . z = 0$

    has a non-trivial solution. For $\lambda = 1$, find all values of α.

15. For what value of k does the following system of equations possess a non - trivial solution over the set of rationales: $3x + ky - 2z = 0$, $x + ky + 3z = 0$, $2x + 3y - 4z = 0$. For that value of k, find all the solutions of the system.

16. If S is the set of distinct values of 'b' for which the following system of linear equations

    $\left.\begin{array}{l} x + y + z = 1 \\ x + ay + z = 1 \\ ax + by + z = 0 \end{array}\right\}$ has no solution, then S is:

    (A) an empty set   (B) An infinite set

    (C) a finite set containing two or more elements (D) A singleton Set

17. Matrix $A = \begin{bmatrix} x & 3 & 2 \\ 1 & y & 4 \\ 2 & 2 & z \end{bmatrix}$, If xyz = 60 and 8x + 4y + 3z = 20, then A(adj A) is equal to:

(A) $\begin{bmatrix} 64 & 0 & 0 \\ 0 & 64 & 0 \\ 0 & 0 & 64 \end{bmatrix}$ (B) $\begin{bmatrix} 88 & 0 & 0 \\ 0 & 88 & 0 \\ 0 & 0 & 88 \end{bmatrix}$ (C) $\begin{bmatrix} 68 & 0 & 0 \\ 0 & 68 & 0 \\ 0 & 0 & 68 \end{bmatrix}$ (D) $\begin{bmatrix} 34 & 0 & 0 \\ 0 & 34 & 0 \\ 0 & 0 & 34 \end{bmatrix}$

18. If n be the number of values of x for which matrix $\Delta(x) = \begin{bmatrix} -x & x & 2 \\ 2 & x & -x \\ x & -2 & -x \end{bmatrix}$ will be singular, then det(Δ(n)) is (where det(B) denotes determinant of Matrix B)

    (A) – 8 (B) – 6 (C) 0 (D) 10

19. A is non-singular square matrix of (n×n) such that $A^2 = -A$, then det (adj (A^n)) is ___

    (A) 1, if n is even (B) - 1, if n is odd (C) 1, if n is odd (D) - 1, if n is even

ANSWER KEY - Matrix & Determinant

SINGLE OPTION CORRECT

1. D
2. D
3. B
4. D
5. D
6. D
7. C
8. D
9. B
10. C
11. B
12. D
13. C
14. A
15. C
16. A
17. C

MULTI OPTION CORRECT

18. A, C
19. A, C
20. A, B, D
21. B, C
22. A, C
23. A, B, C, D
24. C, D
25. B, D
26. A, B, C, D
27. A, B, C, D

INTEGER TYPE

28. 13
29. 0
30. 4
31. 3
32. 4
33. 6
34. 0
35. 3
36. 1
37.
38. 2
39. 1

SUBJECTIVE PROBLEMS

2. (i) $q = 3, p \neq 2$ (ii) $p \neq 2, q \neq 3$ (iii) $p = 2$
3. (i) 8 (ii) $8 \times 40!$
4. 31
5. 8
6. $\dfrac{1}{19}\begin{bmatrix} 48 & -25 \\ -70 & 42 \end{bmatrix}$
7. $x = 2, y = 1, z = -1$
8. (i) $a \neq 3, b \in R$ (ii) $a = -3$ and $b \neq 1/3$ (iii) $a = -3, b = 1/3$
9. 2 or 4
10. 5
11. $\text{Adj}(A) = \begin{bmatrix} 11 & -9 & 1 \\ -7 & 9 & -2 \\ 2 & -3 & 1 \end{bmatrix}$, $A^{-1} = \begin{bmatrix} \dfrac{11}{3} & -3 & \dfrac{1}{3} \\ -\dfrac{7}{3} & 3 & -\dfrac{2}{3} \\ \dfrac{2}{3} & -1 & \dfrac{1}{3} \end{bmatrix}$
13. (i) $\lambda = 3, \mu \neq 10$ (ii) $\lambda \neq 3, \mu \in R$ (iii) $\lambda = 3, \mu = 10$
14. $2\alpha = 2n\pi \pm \dfrac{\pi}{4} + \dfrac{\pi}{4}$
15. $k = 33/2, \ x:y:z = \dfrac{-15}{2} : 1 : -3$
16. D
17. C
18. C
19. A, C

Mathsarc Education

A learning place to fulfill your dream of success!

MATHEMATICS IIT JEE Main/Advanced

RELATIONS & FUNCTIONS

The true test of leadership is how well you function in a crisis.

Section (A) N.C.E.R.T.

1. If $A \times B = \{(a,x), (b,x), (a,y), (b,y)\}$. find A and B.

2. Let $A = \{1, 2\}$ and $B = \{3, 4\}$. Write $A \times B$. How many subsets will $A \times B$ have?

3. Let $A = \{1, 2\}$ and $B = \{1, 2, 3, 4\}$, $C = \{5, 6\}$ and $D = \{5, 6, 7, 8\}$. Verify that

   (i) $A \times (B \cap C) = (A \times B) \cap (A \times C)$ (ii) $A \times C$ is a subset of $B \times D$.

4. Consider $A = \{0, 1, 3, 5\}$. Find the total number of relations on set A.

5. Let $A = \{1, 2, 3, 4\}$, $B = \{1, 5, 9, 11, 15, 16\}$ and $f = \{(1,5), (2,9), (3,1), (4,5), (2,11)\}$. Are the following true?

   (i) f is a relation from A to B   (ii) f is a function from A to B.

   Justify your answer in each case.

6. Select the wrong statements for functions $f: X \to R$ and $g: X \to R$

   Statement 1: $(f + g)(x) = f(x) + g(x)$, $x \in X$.

   Statement 2: $(f - g)(x) = f(x) - g(x)$, $x \in X$

   Statement 3: $(f.g)(x) = f(x).g(x)$, $x \in X$

   Statement 4: $(kf)(x) = k f(x)$, $x \in X$, where k is a real number.

   Statement 5: $\left(\dfrac{f}{g}\right)(x) = \dfrac{f(x)}{g(x)}$, $x \in X, g(x) \neq 0$.

7. Consider the functions $f: D_1 \to R$ and $g: D_2 \to R$, then match the domains for following.

A	$f(x) + g(x)$	P	D_1
B	$f(x) - g(x)$	Q	$D_1 \cap D_2$
C	$f(x) \cdot g(x)$	R	D_2
D	$kf(x)$, where k is a real number.	S	$D_1 \cap D_2 - \{x: g(x)=0\}$
E	$\dfrac{f(x)}{g(x)}$	T	$D_1 \cup D_2$

8. Find the domains of definition of the following functions.

1. $\sin x + \tan x$	2. $\dfrac{\sin x}{\tan x}$	3. $\dfrac{\tan x}{\tan x}$	4. $\sin x \cdot \mathrm{cosec}\, x$
5. $\sin^{-1}(2x^2)$	6. $\dfrac{\sqrt{x^2-4}}{x+2}$	7. $\sqrt{\dfrac{x^2-5x+6}{x+4}}$	8. $\sqrt{x-1} - \sqrt{5-x}$

9. Let $A = \{9,10,11,12,13\}$ and let $f: A \to N$ be defined by $f(n) =$ the highest prime factor of n. Find the range of f.

10. Find the Range of the following functions.

1. $\dfrac{1}{2-\sin 3x}$	2. $|x-2|$	3. $\sqrt{\sin x - 1}$	4. $\dfrac{x^2-1}{x^2+x+1}$
5. $\sin(\cos x)$	6. $3\sin x - 4\cos x$	7. $\dfrac{1}{x^2+1}$	8. $e^{\sin x}$

11. Let $A = R - \{3\}$ and $B = R - \{1\}$. Consider the function $f: A \to B$ defined by $f(x) = \dfrac{x-2}{x-3}$. is f is one-one and onto? Justify your answer.

12. Show that the function $f: R \to R$ defined by $f(x) = 3x^2+5$, is a many one into function.

13. Let $f(x) = \dfrac{x^2-4}{x^2+4}$ for $|x|>2$, then the function $f: (-\infty,-2] \cup [2,\infty) \to (-1,1)$ is

(A) one-one into (B) one-one onto (C) many one into (D) many one onto

14. Let $f: N \to Y$ be a function defined as $f(x) = 4x + 3$, where, $Y = \{y \in N: y = 4x + 3$ for some $x \in N\}$. Show that f is invertible. Find the inverse.

15. If $f(x) = 2x + 5$ and $g(x) = 2x - 5$, $x \in R$, find $(fog)(9)$.

Section (B) JEE-MAINS

1. If $f : (2, 4) \to (1, 3)$ is a function defined by $f(x) = x - \left[\dfrac{x}{2}\right]$ (where [·] denotes the greatest integer function), then $f^{-1}(x)$ is

   (A) $x - 1$ (B) $x + 1$ (C) x (D) none of these

2. If the function $f: [1, \infty) \to (1, \infty)$ is defined by $f(x) = 2^{x(x-1)}$, then $f^{-1}(x)$ is

   (A) $\left(\dfrac{1}{2}\right)^{x(x-1)}$
   (B) $\dfrac{1}{2}\left(1 + \sqrt{1 + 4\log_2 x}\right)$
   (C) $\dfrac{1}{2}\left(1 - \sqrt{1 + 4\log_2 x}\right)$
   (D) not defined

3. Let $h(x) = \dfrac{x}{5^x - 1} + \dfrac{x}{2} + 5$, then $h(x)$ is

   (A) Even
   (B) Odd
   (C) neither even nor odd
   (D) Defined for all $x \in R$

4. Which of the following function is one–one and onto both?

   (A) $f : R \to R, f(x) = e^x$
   (B) $f : R \to R, f(x) = \sin x$
   (C) $f : R^+ \to R, f(x) = \ln x$
   (D) $f : R \to R, f(x) = \tan x$

5. $f(x) = \dfrac{\ln|x|}{\sin^{-1}(x - [x])}$, where [.] = greatest integer function is defined for x belonging to

   (A) R (B) $(0, \infty)$ (C) $R - \{I\}$ (D) none of these

6. The domain of $y = \cos^{-1}\left(\dfrac{1-2|x|}{3}\right) + \log_{|x-1|} x$

   (A) (0, 2) (B) $(0, 1) \cup (1, 2)$ (C) (1, 3) (D) (3, 5)

7. Period of the function $\dfrac{1}{7x - [7x - 8]}$ is, where [.] denotes greatest integer function

   (A) 1 (B) $\dfrac{1}{8}$ (C) $\dfrac{1}{7}$ (D) non periodic

8. If $f(x) = |x-2| + |x-3| + |x-4|$ and $g(x) = f(x+1)$, then $g(x)$ is

   (A) even function (B) odd function

   (C) periodic (D) neither even nor odd

9. If $f(x)$ is a function which is odd and even simultaneously, then $f(3) - f(2)$ is equal to

   (A) 1 (B) –1 (C) 0 (D) none of these

10. The period of $f(x) = 2 + (-1)^{[x]}$ (where [.] denotes greatest integer function) is

    (A) 1.5 (B) 1 (C) 2 (D) 2.5

PARAGRAPH (QUESTION NO. 11, 12, 13)

If $f : X \to Y$ be a function defined by $y = f(x)$ such that f is both one-one and onto then there exists a unique function $g : Y \to X$ such that for each $y \in Y$, $g(y) = x$ iff $y = f(x)$. The function g so defined is called the inverse of f and denoted as $f^{-1}(x) = g(x)$.

11. If $f : R \to R$ be a invertible function such that $f^{-1}(x) = g(x)$ and x_1, x_2 are two distinct roots of the equation

    $f(x) = g(x)$. Then value of $\dfrac{g(x_2) - g(x_1)}{x_2 - x_1}$

    (A) must be –1 (B) must be 1 (C) may be 3 (D) cannot determine

12. If $f : \left[\dfrac{3\pi}{2}, 2\pi\right] \to [-1, 0]$, where $f(x) = \sin x$ then $f^{-1}(x)$ is

    (A) $\dfrac{3\pi}{2} + \sin^{-1} x$ (B) $2\pi + \sin^{-1} x$ (C) $3\pi - \sin^{-1} x$ (D) $\dfrac{5\pi}{2} - \sin^{-1} x$

13. A function $f : I \to J$ given by $f(i) = j$ where $I = \{0,1,2,....9\}$, $J = \{0,1,2, ...100\}$ and i, j are element of set I, J respectively. Then number of bijective functions of type $f: I \to B$ where $B \subseteq J$ and $f(5) = 5$ is

   (A) $^{100}C_9 10!$ (B) $^{100}C_9 .9!$ (C) $^{101}C_{10} 10!$ (D) none of these

14. If $f(x)$ is differentiable and $\int_0^{t^2} x f(x) dx = \frac{2}{5} t^5$, then $f\left(\frac{4}{25}\right)$ equals

   (A) 1 (B) –5/2 (C) 2/5 (D) 5/2

15. Find the domain and range of $h(x) = g(f(x))$, where

   $f(x) = \begin{cases} [x], & -2 \le x \le -1 \\ |x|+1, & -1 < x \le 2 \end{cases}$ and $g(x) = \begin{cases} [x], & -\pi \le x \le 0 \\ \sin x, & 0 < x \le \pi \end{cases}$, and $[.] = $ G.I.F.

Section (C) JEE-ADVANCE

1. Find the range of the function $f(x) = \log_2\left(2 - [\log_{\sqrt{2}}(16\sin^2 x + 1)]\right)$; where [.] denotes the greatest integer function.

2. If $f(x) = \cos^{-1}\left(\frac{2-|x|}{4}\right) + \{\log(3 - x)\}^{-1}$, then find its domain.

3. A function $f: R \to R$ is defined such that $f(x + y) = f(x) + 2y^2 + kxy$, $\forall x, y \in R$ and $f(1) = 2$, $f(2) = 8$, show that $f(x + y) \cdot f\left(\frac{1}{x+y}\right) = k$, for $x + y \ne 0$ Hence (or otherwise) find $f(x)$.

4. Let $f(x) = \frac{\alpha x}{x+1}$, $x \ne -1$, for what values of α, is $f(f(x)) = x$

   (A) $\sqrt{2}$ (B) $-\sqrt{2}$ (C) 1 (D) –1

5. The period of the function $f(x) = 3\sin\frac{\pi x}{3} + 4\cos\frac{\pi x}{4}$ is

   (A) 6 (B) 24 (C) 8 (D) 2π

6. Verify whether function $f(x) = \dfrac{2x(\sin x + \tan x)}{2\left[\dfrac{x+2\pi}{\pi}\right] - 3}$, $(x \neq n\pi)$ is even or odd function. (where [.] is the greatest integer function.)

7. A real valued function satisfying the functional equation $2f(\sin x) + f(\cos x) = x$, then prove the following results.

   (a) Domain of $f(x)$ is $[-1, 1]$

   (b) Range of $f(x)$ is $\left[-\dfrac{2\pi}{3}, \dfrac{\pi}{3}\right]$

8. (a) If the function $f(x) = x + e^{\left(\frac{x}{2}\right)+1}$, find $3(f^{-1}(-1))'$

   (b) Number of solutions of the equation $1 = |\ln|x|| \, 4^{|x|}$

9. Let $f: R \to R$ be a continuous & differentiable function given by $f(x) = x + \int_0^1 (xy + x^2) f(y) dy$. Then

   (A) $\int_0^1 f(x)dx = \dfrac{26}{23}$
   (B) $\int_0^1 f(x)dx = \dfrac{25}{13}$
   (C) $\int_0^1 xf(x)dx = \dfrac{13}{25}$
   (D) $\int_0^1 xf(x)dx = \dfrac{25}{23}$

10. Let f be a real valued function defined on the interval $(-1, 1)$ such that $f(x) = e^{2x} \int_0^x \sqrt{1+t^3} \, dt$ and Let f^{-1} be the inverse of f then $\dfrac{d}{dx}\left(f^{-1}(x)\right)$ at $x = 0$ is equal to

11. Let $f: \left[\dfrac{-\pi}{3}, \dfrac{k\pi}{3}\right] \to [-1, 2]$ where $f(x) = \sqrt{3} \sin x - \cos x + 1$. The value of k for which $f(x)$ becomes invertible function.

    (A) 1   (B) 2   (C) 3/2   (D) 1/2

12. Let $f: R \to R$ be a differentiable function and $f(1) = 4$. Then the value of $\lim\limits_{x \to 1} \int_4^{f(x)} \dfrac{2t}{x-1} dt$ is

    (A) $8f'(1)$ (B) $4f'(1)$ (C) $2f'(1)$ (D) $f'(1)$

13. Maximum value of parameter a for which $\exists$ a real number x satisfying $\sqrt{1-x^2} \geq a - \sqrt{3} \, x$.

14. The function f : [0, 3] → [1, 29], defined by $f(x) = 2x^3 - 15x^2 + 36x + 1$, is

   (A) one-one and onto (B) onto but not one-one

   (C) one-one but not onto (D) neither one-one nor onto

15. Let f : (0, 1) → R be defined by $f(x) = \dfrac{b-x}{1-bx}$. Where b is a constant such that 0 < b < 1. Then

   (A) f is not invertible on (0, 1) (B) $f \neq f^{-1}$ on (0, 1) and $f'(b) = \dfrac{1}{f'(0)}$

   (C) $f = f^{-1}$ on (0, 1) and $f'(b) = \dfrac{1}{f'(0)}$ (D) f^{-1} is differentiable on (0, 1)

Mathsarc Education

A learning place to fulfill your dream of success!

MATHEMATICS   IIT JEE Main/Advanced

DEFINITE INTEGRATION

Delta increment in learning leads to integrate your knowledge, required for success!

SINGLE OPTION CORRECT

1. The value of $\int_{0}^{\pi/2} \sin 8x \cot x \, dx + \int_{-\pi/4}^{\pi/4} \ln\left(\dfrac{1-\sin x}{1+\sin x}\right) dx$ is equal to

    (A) 0   (B) 1   (C) $\pi/4$ (D) $\pi/2$

2. Let R be the set of all real numbers and f: R→ R be defined as $f(x) = \int_{0}^{1} |x - t| \, dt$. The number of times the graph of g(x) = f(x) – x cuts the x – axis between the lines x = 0 and x = 1, is

    (A) 0   (B) 1   (C) 2   (D) 3

3. The value of $\int_{0}^{\pi} x\sqrt{1+|\cos x|} \, dx$ is equal to

    (A) $2\sqrt{2}\,\pi$ (B) $\sqrt{2}\,\pi$ (C) 2π (D) 4π

4. Let $f(x) = \int_{0}^{x} \dfrac{dt}{\sqrt{1+t^3}}$ and g(x) be the inverse of f(x), then which one of the following holds good?

    (A) $2g'' = g^2$ (B) $2g'' = 3g^2$ (C) $3g'' = 2g^2$ (D) $3g'' = g^2$

5. If $g(x) = \int_{1}^{x} e^{t^2} dt$ then the value of $\int_{3}^{x^3} e^{t^2} dt$ equals

    (A) $g(x^3) - g(3)$ (B) $g(x^3) + g(3)$ (C) $g(x^3) - 3$ (D) $g(x^3) - 3g(x)$

6. Let f(x) be a continuous function such that its first two derivatives f'(x), f"(x) are continuous. The tangents to the graph of f(x) at the points with abscissa x = a and x = b make with the X-

axis angles $\frac{\pi}{3}$ & $\frac{\pi}{4}$ respectively. Then the value of the integral $\int_a^b f'(x)f''(x)dx$ equals

(A) $1 - \sqrt{3}$ (B) 0 (C) 1 (D) -1

7. If $S = \int_0^1 \frac{e^t}{t+1}dt$ then $S_1 = \int_{a-1}^a \frac{e^{-t}}{t-a-1}dt$ is

(A) Se^a (B) Se^{-a} (C) $-Se^{-a}$ (D) $-Se^a$

8. The value of the integral $\int_0^\pi |1+2\cos x|dx$ is

(A) $\frac{\pi}{3}+\sqrt{3}$ (B) $\frac{\pi}{3}+2\sqrt{3}$ (C) $\frac{\pi}{3}+4\sqrt{3}$ (D) $\frac{2\pi}{3}+4\sqrt{3}$

9. The value of the integral $\int_0^u \sqrt{1+\sin\left(\frac{x}{2}\right)}dx$, where $0 \leq u \leq \pi$, is

(A) $4+4\left(\sin\frac{u}{4}-\cos\frac{u}{4}\right)$ (B) $4-4\left(\sin\frac{u}{4}-\cos\frac{u}{4}\right)$

(C) $4+\frac{1}{4}\left(\cos\frac{u}{4}-\sin\frac{u}{4}\right)$ (D) $4-\frac{1}{4}\left(\cos\frac{u}{4}-\sin\frac{u}{4}\right)$

10. The definite integral $\int_0^{\pi/2} \frac{dx}{1+(\tan x)^{101}}$ equal

(A) π (B) $\pi/2$ (C) 0 (D) $\pi/4$

11. The value of the integral $\int_0^{\pi/4} \log_e(1+\tan\theta)d\theta$ is

(A) $\pi/8$ (B) $\frac{\pi}{8}\log_e 2$ (C) 1 (D) $2\log_e 2 - 1$

12. Define the real-valued function f on the set of real numbers by $f(x) = \int_0^1 \frac{x^2+t^2}{2-t}dt$. Consider the curve $y = f(x)$. It represents

(A) a straight line (B) a parabola (C) a hyperbola (D) an ellipse

13. The value of $\lim_{n\to\infty} \dfrac{1}{n} \sum_{r=0}^{n-1} \cos\left(\dfrac{r\pi}{2n}\right)$

    (A) 1   (B) 0   (C) $2/\pi$ (D) does not exist

14. $\lim_{n\to\infty} \dfrac{\sqrt{1}+\sqrt{2}+.....+\sqrt{n-1}}{n\sqrt{n}}$ is equal to

    (A) 1/2   (B) 1/3   (C) 2/3   (D) 0

15. The value of $\lim_{n\to\infty} \sum_{i=1}^{n} \dfrac{1}{n}\left[\sqrt{4i/n}\right]$, where [.] = GIF, is

    (A) 3   (B) 3/4   (C) 4/3   (D) None of these

16. $\int_{-1}^{3/2} |x\sin\pi x|\, dx$ is equal to

    (A) $\dfrac{3\pi+1}{\pi^2}$ (B) $\dfrac{\pi+1}{\pi^2}$ (C) $\dfrac{1}{\pi^2}$ (D) $\dfrac{3\pi-1}{\pi^2}$

17. Let $f(x)=\int_{0}^{x} e^{-t^2}\, dt, \forall x > 0$. Then for all $x > 0$

    (A) $xe^{-x^2} < f(x)$ (B) $x < f(x)$ (C) $1 < f(x)$ (D) None of these

18. The value of the integral $\int_{0}^{\log_e 5} \dfrac{e^x\sqrt{e^x-1}}{e^x+3}\, dx$ is

    (A) 4π (B) 4   (C) $\pi/2$ (D) $4-\pi$

19. The value of $\int_{-\pi/2}^{\pi/2} e^{-x^2/2} \sin x\, dx$ is

    (A) $\dfrac{\pi}{2}-1$ (B) $\dfrac{\pi}{2}$ (C) $\sqrt{2\pi}$ (D) None of these

20. Let $g(t) = \int_{-10}^{t} (x^2+1)^{10}\, dx$ for all $t \geq -10$. Then

    (A) g is not differentiable
    (B) g is constant
    (C) g is increasing in $(-10, \infty)$
    (D) g is decreasing in $(-10, \infty)$

21. If $f(x) = \cos x - \int_0^x (x-t)f(t)dt$, then $f''(x) + f(x)$ equals

(A) $-\cos x$ (B) 0   (C) $\int_0^x (x-t)f(t)dt$ (D) $-\int_0^{-x}(x-t)f(t)dt$

22. If $I_n = \int_0^{\pi/2} x^n \cos x\, dx$, then the value of $2^8(I_8 + 56 I_6)$ is

(A) π^8 (B) 8^π (C) 5^π (D) π^5

23. $\int_0^\pi e^{\sin x} \sin 2x \cdot \sin^3((2n+1)x)\, dx$ is equal to ($\forall\, n \in I$)

(A) 0   (B) 1   (C) $\pi/2$ (D) π

24. Let $I_n = \int_0^\pi \dfrac{\sin\left(n+\frac{1}{2}\right)x}{\sin\left(\frac{x}{2}\right)}\, dx$. The value of $I_1^2 + I_2^2 + I_3^2 + \ldots + I_{20}^2$ is

(A) $20\pi^2$ (B) $10\pi^2$ (C) $5\pi^2$ (D) $30\pi^2$

25. The complete set of values of the parameter 'a' such that $\int_0^{2a} (|x-a|+5)\, dx > 24$

(A) $(-\infty, -12)$ (B) $(-\infty, -12) \cup (2, \infty)$ (C) $(2, \infty)$ (D) None of these

26. Number of solutions of the equation $6\int_0^{|x|}((t^2-1)\ln t)\, dt = 5|x|, x \in R_0$ is

(A) 5   (B) 4   (C) 2   (D) 3

27. If $f(x) = \dfrac{\sin x}{x}$, $I = \int_0^\pi f(x)\, dx$ & $J = \int_0^{\pi/4} f(x) f\left(\dfrac{\pi}{2}-x\right)dx$ then $\dfrac{I}{J}$ is equal to

(A) 2π (B) $-\pi$ (C) -2π (D) π

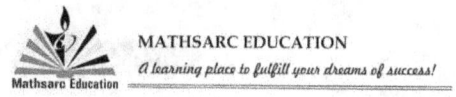

MULTIPLE OPTIONS CORRECT

1. If z is an unimodular complex number such that $\text{Re}(z-3)+\text{Re}(z^2) = \int_0^{\pi/2} \sin x \ln|\sin x - \cos x| dx$ then which of the following can be true?

   (A) $z+\bar{z}=1$ (B) $z+\bar{z}=-2$ (C) $\arg(z)=\dfrac{\pi}{3}$ (D) $\arg(z) = \pi$

2. If the value of $\left[\dfrac{\left|\int_{-100}^{100}[t^3]dt\right|}{12}\right]$ is equal to I, then I satisfies (where [.] = G.I.F.)

   (A) $I \leq 8$ (B) $I \leq 7$ (C) $I \geq 2$ (D) $I \geq 5$

3. Let $f(x) = \lim_{n \to \infty} \dfrac{\cos x}{1+(\tan^{-1}x)^n}$ and let $I = \int_0^\infty f(x)dx$.

   (A) $I > \tan 1$ (B) $I < \tan 1$ (C) $I > \sin 1$ (D) $I < \sin 1$

4. If f is differentiable function such that $f(x + y) = f(x) \cdot f(y) \; \forall \; x, y \in R$, $f(0) \neq 0$ and $f'(0) = 1$. Then

   (A) $\int_{-1}^{1} \dfrac{\ln(f(x))^2}{1+(f(x))^2} dx = \dfrac{1}{3}$ (B) $\int_{-\pi/2}^{\pi/2} \dfrac{\cos(\ln(f(x)))}{1+f(x)} dx = 1$

   (C) $\int_{-2017}^{2017} \ln\left(\ln(f(x)) + \sqrt{1+(\ln(f(x)))^2}\right) dx = 0$ (D) $\int_0^2 (f(x) - \ln(f(x))) dx = e^2 - 2$

5. Consider $L = 3\sqrt{12} + 3\sqrt{13} + 3\sqrt{14} + ... + 3\sqrt{1011}$, $R = 3\sqrt{13} + 3\sqrt{14} + ... + 3\sqrt{1012}$. If $I = \int_{12}^{1012} 3\sqrt{x}\, dx$ then which of the following is/are correct?

   (A) $L + R < 2I$ (B) $R > I$ (C) $L < I$ (D) $L + R > 2I$

6. If $I_n = \int_{-\pi}^{\pi} \dfrac{\sin nx}{(1+\pi^x)\sin x}dx, n = 0, 1, 2, 3......$, then

   (A) $I_n = I_{n+2}$ (B) $\sum_{m=1}^{10} I_{2m+1} = 10\pi$ (C) $\sum_{m=1}^{10} I_{2m} = 0$ (D) $I_n = I_{n+1}$

7. Let $\sum_{k=1}^{89} \cos^6(k°) = A$, then -

   (A) $A > \int_0^{\pi/2} \cos^6 \theta d\theta$ (B) $A \in [20, 30]$

   (C) $A < \dfrac{180}{\pi} \int_0^{\pi/2} \cos^6 \theta d\theta$ (D) $A < \int_0^{\pi/2} \cos^6 \theta d\theta$

8. Let f(x) is a differentiable function such that $f'(x) = f(x) + \int_0^2 f(x)dx$ and $f(0) = \dfrac{4-e^2}{3}$.

   (A) The number of solutions of x + f(x) = 0 is 2.   (B) The number of solutions of x + f(x) = 0 is 1.

   (C) $\int_0^1 f(x)dx = \dfrac{3e - e^2 - 2}{3}$ (D) $\int_0^1 f(x)dx = \dfrac{3e + e^2 - 4}{3}$

INTEGER TYPE

1. Let $I_n = \int_0^{\pi/4} \tan^n x\, dx$ where n = 0, 1, 2, 3,..... and $S_n = \sum_{n=0}^{n}(I_n I_{n+1} + I_n I_{n+3} + I_{n+2} I_{n+1} + I_{n+2} I_{n+3})$. Find the value of $\lim_{n\to\infty} S_n$.

2. If a > 2 and $\int_0^\infty \dfrac{dx}{a^2 + \left(x - \dfrac{1}{x}\right)^2} = \dfrac{\pi}{5050}$ and a = 101 α², then |α| =

3. Let $f : D \to y$ defined as $f(x) = \ln\left[\cos|x| + \dfrac{1}{2}\right]$ (where [.] represents greatest integer function), then $\int_{x_1}^{x_2}\left(\lim_{n\to\infty}\dfrac{(f(x))^n}{x^2 + \tan^2 x}\right)dx$ is equal to............. (where $x_1, x_2 \in D$)

4. Let $I_1 = \int_0^1 \dfrac{e^x}{1+x}dx$ & $I_2 = \int_0^1 \dfrac{x^2 dx}{e^{x^3}(2-x^3)}$. What is the value of $\dfrac{I_1}{e \cdot I_2}$?

5. If 'a' is a positive integer, then the number of solutions of inequation
$\int_0^{\pi/2} \left\{ a^2\left[\dfrac{\cos(3x)}{4} + \dfrac{3}{4}\cos x\right] + a\sin x - 20\cos x \right\}dx \leq -\dfrac{a^2}{3}$ is _____

6. If f is a differentiable function such that $f(f(x)) = x \; \forall \; x \in [0, 1]$.
The value of $2017 \int_0^1 (x - f(x))^{2016} dx$, if $f(0) = 1$.

MATRIX MATCH

1. Match the following

	COULUMN - I		COULUMN - II
A	$\int_{-1}^{1} \left(\dfrac{d}{dx}\left(\dfrac{1}{1+e^{1/x}}\right)\right)dx = \dfrac{k}{1+e}$, then k is	P	1
B	$\int_0^{\infty} \dfrac{dx}{\left(x+\sqrt{1+x^2}\right)^3} = k$ then value of 8k is	Q	2
C	$\displaystyle\lim_{n\to\infty} n \int_0^{\pi/2} \left((\sin x)^{1/n} - 1\right)dx = -\dfrac{\pi \cdot \ln k}{k}$ then k is	R	3
D	$\dfrac{\int_0^1 (1-x^7)^{1/3} dx}{\int_0^1 (1-x^3)^{1/7} dx} = k$ then k is	S	4

THANKS!

Keep smiling!

Visit Us: https://www.mathsarc.com

ANSWER KEY – Definite Integration

SINGLE OPTION CORRECT

1. D
2. B
3. C
4. B
5. A
6. D
7. C
8. B
9. A
10. D
11. B
12. B
13. C
14. C
15. A
16. A
17. A
18. D
19. D
20. C
21. A
22. A
23.
24. A
25. B
26. B
27. D

MULTI OPTIONS CORRECT

1.
2.
3.
4.
5.
6.
7. A, B, C
8. B, C

INTEGER TYPE

1. 1
2. 5
3. 0
4. 3
5. 4
6. 1

Mathsarc Education

A learning place to fulfill your dream of success!

MATHEMATICS   IIT JEE Main/Advanced

VECTOR & 3D

Follow your dream, you will definitely contribute in human life!

SINGLE OPTION CORRECT

1. The vector $x\hat{i} + y\hat{j} + z\hat{k}$ makes with the plane of the two vectors (2, 3, -1) and (1, -1, 2) an acute angle $\cot^{-1}\sqrt{2}$. Then

   (A) $y(x + z) = zx$ (B) $z(x + y) = xy$ (C) $x(y + z) = yz$ (D) $(x + y + z) = xyz$

2. Let $\vec{b} = -\hat{i} + 4\hat{j} + 6\hat{k}$ and $\vec{c} = 2\hat{i} - 7\hat{j} - 10\hat{k}$. If $\vec{a}$ be a unit vector and the scalar triple product $\begin{bmatrix} \vec{a} & \vec{b} & \vec{c} \end{bmatrix}$ has the greatest value, then $\vec{a}$ is equal to

   (A) $\dfrac{1}{\sqrt{3}}(\hat{i} + \hat{j} + \hat{k})$

   (B) $\dfrac{1}{\sqrt{5}}(\sqrt{2}\hat{i} - \hat{j} - \sqrt{2}\hat{k})$

   (C) $\dfrac{1}{3}(2\hat{i} + 2\hat{j} - \hat{k})$

   (D) $\dfrac{1}{\sqrt{59}}(3\hat{i} - 7\hat{j} - \hat{k})$

3. If (α, β, γ) be intersection point of lines $x - 3y + 2z + 4 = 0 = 2x + y + 4z + 1$ & $\dfrac{x - \frac{1}{3}}{8} = \dfrac{y}{3} = \dfrac{z}{-6}$, then $(\alpha + \beta + \gamma)$ is-

   (A) -2 (B) -1 (C) 0 (D) 2

4. If $\vec{r} = 3\hat{i} + 2\hat{j} - 5\hat{k}$, $\vec{a} = 2\hat{i} - \hat{j} + \hat{k}$, $\vec{b} = \hat{i} + 3\hat{j} - 2\hat{k}$ and $\vec{c} = -2\hat{i} + \hat{j} - 3\hat{k}$ such that $\vec{r} = \lambda\vec{a} + \mu\vec{b} + \gamma\vec{c}$, then-

   (A) $\mu, \dfrac{\lambda}{2}, \gamma$ are in A.P.

   (B) $2\mu, \lambda, \gamma$ are in A.P.

   (C) μ, λ, γ are in A.P.

   (D) $\lambda, \dfrac{\mu}{3}, \gamma$ are in G.P.

5. If a, b, c are pth, qth, rth terms of an H.P. and $\vec{u} = (q-r)\hat{i} + (r-p)\hat{j} + (p-q)\hat{k}$, $\vec{v} = \dfrac{\hat{i}}{a} + \dfrac{\hat{j}}{b} + \dfrac{\hat{k}}{c}$, then:

   (A) $\vec{u}, \vec{v}$ are parallel vectors

   (B) $\vec{u}, \vec{v}$ are orthogonal vectors

   (C) $\vec{u} \cdot \vec{v} = 1$

   (D) $\vec{u} \times \vec{v} = \hat{i} + \hat{j} + \hat{k}$

6. The image of the point (1, 2, –1), on the plane containing the line $\dfrac{x+1}{-3} = \dfrac{y-3}{2} = \dfrac{z+2}{1}$ and the point $(0, 7, -7)$ is.

   (A) $\left(\dfrac{-1}{3}, -\dfrac{7}{3}, \dfrac{1}{3}\right)$
   (B) $\left(\dfrac{-1}{3}, \dfrac{2}{3}, \dfrac{-7}{3}\right)$
   (C) $\left(\dfrac{-1}{3}, 0, \dfrac{-7}{3}\right)$
   (D) $\left(\dfrac{-1}{3}, \dfrac{2}{3}, \dfrac{7}{3}\right)$

7. $(\vec{a} \times \vec{b}) \times \left[(\vec{b} \times \vec{c}) \times (\vec{a} \times \vec{b} + \vec{b} \times \vec{c} + \vec{c} \times \vec{a})\right]$ is

   (A) $\begin{bmatrix} \vec{a} & \vec{b} & \vec{c} \end{bmatrix} \left[(\vec{b} \cdot \vec{a} + \vec{a} \cdot \vec{c})\vec{b} - (|\vec{b}|^2 + \vec{b} \cdot \vec{c})\vec{a} \right]$

   (B) $\begin{bmatrix} \vec{a} & \vec{b} & \vec{c} \end{bmatrix} \left[(\vec{b} \cdot \vec{a} + \vec{a} \cdot \vec{c})\vec{b} + (|\vec{b}|^2 - \vec{b} \cdot \vec{c})\vec{a} \right]$

   (C) $\begin{bmatrix} \vec{a} & \vec{b} & \vec{c} \end{bmatrix} \left[(\vec{b} \cdot \vec{a} - \vec{a} \cdot \vec{c})\vec{b} + (|\vec{b}|^2 + \vec{b} \cdot \vec{c})\vec{a} \right]$

   (D) $\begin{bmatrix} \vec{a} & \vec{b} & \vec{c} \end{bmatrix} \left[(\vec{a} \cdot \vec{c} - \vec{b} \cdot \vec{a})\vec{b} + (|\vec{b}|^2 - \vec{b} \cdot \vec{c})\vec{a} \right]$

8. Let $\vec{V} = 2\hat{i} + \hat{j} - \hat{k}$, $\vec{W} = \hat{i} + 3\hat{k}$, $|\vec{U}| = 2$. If $\vec{U}$ is a vector in x-y plane, then greatest value of $\begin{bmatrix} \vec{U} & \vec{V} & \vec{W} \end{bmatrix}^2$ is -

   (A) 232
   (B) 340
   (C) 236
   (D) 312

9. The line L given by $\dfrac{x-2}{2} = \dfrac{y-1}{b} = \dfrac{z+1}{c}$ passes through the point (1, 2, 3). Another line K is parallel to line L and has the equation $\dfrac{x+2}{a} = \dfrac{y-3}{2} = \dfrac{z+4}{d}$. Then the distance between line L and K is

   (A) $\dfrac{\sqrt{297}}{3}$
   (B) $\dfrac{\sqrt{243}}{3}$
   (C) $\dfrac{\sqrt{272}}{9}$
   (D) $\dfrac{\sqrt{291}}{9}$

10. If the plane 2x + y + 2z = 9 intersects the co-ordinate axes in A, B and C and the co-ordinates of orthocenter of triangle ABC be (α, β, γ), then the value of α + β + γ is

    (A) 1
    (B) 3
    (C) 5
    (D) 7

11. If lines $\dfrac{x-3}{2}=\dfrac{y+1}{-3}=\dfrac{z+a}{p}$ and $\dfrac{x+2}{2}=\dfrac{y-4}{4}=\dfrac{z+5}{2}$ are perpendicular coplanar lines, then value of a + p is -

   (A) 3/5 (B) - 2/7 (C) 4/7 (D) - 3/5

12. If the vectors $(1-x)\hat{i}+\hat{j}+\hat{k}, \hat{i}+(1-y)\hat{j}+\hat{k}$ and $\hat{i}+\hat{j}+(1-z)\hat{k}$ are coplanar vectors, then value of $\dfrac{1}{x}+\dfrac{1}{y}+\dfrac{1}{z}$ is (x, y, z are non-zero numbers)

   (A) 0 (B) 3 (C) 1/3 (D) 1

13. Equation of line of shortest distance between the lines $\dfrac{x}{2}=\dfrac{y}{-3}=\dfrac{z}{1}$ & $\dfrac{x-2}{3}=\dfrac{y-1}{-5}=\dfrac{z+2}{2}$ is -

   (A) 3(x – 21) = (3y – 92) = (3z – 32)
   (B) 3x – 62 = 3y – 93 = 3z + 31
   (C) $\dfrac{(x-21)}{3}=\dfrac{\left(y+\dfrac{92}{3}\right)}{3}=\dfrac{\left(z-\dfrac{32}{3}\right)}{3}$
   (D) $x-\dfrac{62}{3}=y+31=\left(z+\dfrac{31}{3}\right)$

14. A variable plane passes through the point (1, 2, 3) and meets the coordinate axis is P, Q, R. Then, the locus of the point common to the planes through P, Q, R parallel to coordinate planes

   (A) contains point (3, 6, 9)
   (B) passes through (0, 0, 0)
   (C) is $\dfrac{1}{x}+\dfrac{1}{y}+\dfrac{1}{z}=1$
   (D) contains line $\dfrac{x-1}{2}=\dfrac{y-2}{4}=\dfrac{z+6}{15}$

15. If $\vec{a}, \vec{b}, \vec{c}$ are three non-coplanar unit vectors, the angle between them pair wise are $\dfrac{\pi}{6}, \dfrac{\pi}{4}$ & $\dfrac{\pi}{3}$, then the value of $\left|\begin{bmatrix}\vec{a} & \vec{b} & \vec{c}\end{bmatrix}\right|$ is

   (A) $\dfrac{\sqrt{3}-1}{2\sqrt{2}}$ (B) $\dfrac{\sqrt{3}+1}{2\sqrt{2}}$ (C) $\dfrac{\sqrt{2+\sqrt{6}}}{2}$ (D) $\dfrac{\sqrt{\sqrt{6}-2}}{2}$

16. A plane containing the line $\dfrac{x-3}{2}=\dfrac{y-1}{4}=\dfrac{z-2}{5}$ and it is passing through the point (4, 3, 7). The equation of the plane is

   (A) 4x + 8y + 8z = 4 (B) 4x – 8y – 8z = 4 (C) 4x – 2y – 10 = 0 (D) 4x – 8y + 8z = 4

17. If vectors $\hat{a}, \hat{b}, \hat{c}$ are in space such that $\hat{a}\cdot\hat{b} = \hat{b}\cdot\hat{c} = \hat{c}\cdot\hat{a} = \dfrac{1}{2}$, then $(\hat{a}\times\hat{b})\cdot(\hat{a}\times(\hat{b}\times\hat{c}))$ is equal to

    (A) $\dfrac{1}{\sqrt{2}}$ 　　　　(B) $-\dfrac{1}{\sqrt{2}}$ 　　　　(C) $\dfrac{1}{2\sqrt{2}}$ 　　　　(D) $-\dfrac{1}{2\sqrt{2}}$

18. Equation of plane perpendicular to x = 0 and passing through (2, −1, 0) can be

    (A) x + 3y + z + 1 = 0　　(B) 2y − z + 2 = 0　　(C) y = − 1　　(D) 2y + z + 1 = 0

19. Let P(x, y, 1) and Q(x, y, z) lies on the curve $\dfrac{x^2}{9} + \dfrac{y^2}{4} = 4$ and $\dfrac{x+2}{1} = \dfrac{\sqrt{3}-y}{\sqrt{3}} = \dfrac{z-1}{2}$ respectively. Then minimum distance between P and Q is

    (A) √2　　　　(B) $\sqrt{\dfrac{7}{2}}$　　　　(C) 2　　　　(D) None of these

20. The value of 'a' so that the volume of parallelepiped formed by $\hat{i}+\hat{j}+\hat{k}, \hat{j}+a\hat{k}$ and $a\hat{i}+\hat{k}$ becomes minimum is

    (A) 3　　　　(B) − 1/3　　　　(C) 1/2　　　　(D) − 2

21. Let $\vec{a}, \vec{b}\ \&\ \vec{c}$ be non-zero vectors such that $(\vec{a}\times\vec{b})\times\vec{c} = -\dfrac{1}{3}|\vec{b}||\vec{c}|\vec{a}$. If θ is the acute angle between the vectors $\vec{b}\ \&\ \vec{c}$, then sinθ equal to:

    (A) 1/3　　　　(B) √2/3　　　　(C) 2/3　　　　(D) 2√2/3

22. A square ABCD of a diagonal 2a is folded along the diagonal AC, so that the planes DAC, BAC are at right angle. The shortest distance between DC and AB is

    (A) $\sqrt{\dfrac{2}{3}}a$　　　　(B) $\dfrac{2a}{\sqrt{3}}$　　　　(C) $\sqrt{\dfrac{2a}{3}}$　　　　(D) None of these

23. If the plane passing through the points (λ,1,1), (1,2,1) and (2,3,4) is parallel to the line $\vec{r} = \mu(\hat{i}+\hat{j}+2\hat{k}), (\mu \in R)$, then λ is equal to-

    (A) − 1/2　　　　(B) − 1　　　　(C) 3/2　　　　(D) 0

24. If for unit vectors $\hat{a}, \hat{b}$ and non-zero $\vec{c}$, $\hat{a}\times\hat{b}+\hat{a} = \vec{c}$ and $\hat{b}\cdot\vec{c} = 0$, then volume of parallelepiped with coterminous edges $\hat{a}, \hat{b}$ and $\vec{c}$ will be (in cu.units)-

    (A) 6　　　　(B) 4　　　　(C) 1　　　　(D) 1/2

MULTIPLE OPTIONS CORRECT

1. Let $\vec{A} = a\hat{i} + b\hat{j} + c\hat{k}$ be a unit vector and $\vec{B}$ is another vector in R³ such that $|\vec{A} \times \vec{B}| = 1$, $\vec{C} = \frac{1}{3}(2\hat{i} + 2\hat{j} - \hat{k})$ and $(\vec{A} \times \vec{B}) \cdot \vec{C} = 1$, then which of the following statement(s) is (are) correct?

   (A) If $\vec{A}$ lies in plane x + y + z = 10, then there are exactly 2 choices for $\vec{A}$.

   (B) If $\vec{A}$ lies in plane x + y + z = 10, then there are exactly 4 choices for $\vec{A}$.

   (C) If a, b, c ∈ I, then there is no such vector $\vec{A}$.

   (D) If a, b, c ∈ I, then there are infinitely many choices for $\vec{A}$.

2. Consider a variable plane lx + my + nz = k (k > 0) and l, m, n are direction cosines of normal of the plane. Let the given plane intersects co-ordinate axis at A, B, C, then area of ΔABC may be -

   (A) $\frac{3\sqrt{3}k^2}{2}$ (B) $\frac{3\sqrt{3}k^2}{4}$ (C) $3\sqrt{3}k^2$ (D) $12\sqrt{3}k^2$

3. Consider the lines, $L_1 : \frac{x}{2} = \frac{y}{-3} = \frac{z}{1}$ & $L_2 : \frac{x-2}{3} = \frac{y-1}{-5} = \frac{z+2}{2}$, then the line along shortest distance can be, constituted by the line of intersection of planes

   (A) 4x + y − 5z = 0 (B) x − 3y + 5z = 0 (C) 5x − 7y + 2z = −1 (D) 7x + y − 8z = 31

4. $\vec{p}, \vec{q}, \vec{r}$ be vectors such that $\vec{q} \cdot \vec{r} = 0$ and $\vec{p} \cdot \vec{q} \neq 0$. Let α is real constant such that $\vec{x} \cdot \vec{p} = \alpha$; $\vec{x} \times \vec{q} = \vec{r}$, then $\vec{x} = \lambda_1 \vec{q} + \lambda_2 (\vec{p} \times \vec{r})$ where

   (A) $\lambda_1 = \frac{\alpha}{\vec{p} \cdot \vec{q}}$ (B) $\lambda_2 = \frac{1}{\vec{p} \cdot \vec{q}}$ (C) $\lambda_2 = \frac{1}{\vec{r} \cdot \vec{q}}$ (D) $\lambda_1 = \frac{\alpha}{\vec{r} \cdot \vec{q}}$

5. Let $\hat{u}, \hat{v}, \hat{w}$ be three unit vectors such that $\hat{u} + \hat{v} + \hat{w} = \hat{a}$, $\hat{a} \cdot \hat{u} = \frac{3}{2}$, $\hat{a} \cdot \hat{v} = \frac{7}{4}$ & $|\hat{a}| = 2$, then

   (A) $\hat{v} \cdot \hat{u} = \frac{3}{4}$ (B) $\hat{u} \cdot \hat{w} = 0$ (C) $\hat{u} \cdot \hat{v} = \frac{1}{4}$ (D) $\hat{u} \cdot \hat{w} = -\frac{1}{4}$

6. $\vec{a}$ & $\vec{b}$ be two vectors such that $|\vec{a}|=1, |\vec{b}|=4$ & $\vec{a}\cdot\vec{b}=2$. If $\vec{c}=(2\vec{a}\times\vec{b})-3\vec{b}$, then which of the following is/are correct?

   (A) $\vec{b}\cdot\vec{c}=48$

   (B) $\vec{b}\cdot\vec{c}=-48$

   (C) Angle between $\vec{b}$ & $\vec{c}$ is $\dfrac{5\pi}{6}$

   (D) Angle between $\vec{b}$ & $\vec{c}$ is $\dfrac{\pi}{6}$

INTEGER TYPE

1. A plane P intersects 4 lines L_1, L_2, L_3 and L_4 (given below) at A, B, C, D.
$L_1: \dfrac{x-3}{2}=\dfrac{y-3}{1}=\dfrac{z-3}{2}$, $L_2: \dfrac{x-3}{2}=\dfrac{y-3}{1}=\dfrac{z}{2}$, $L_3: \dfrac{x}{2}=\dfrac{y-3}{1}=\dfrac{z}{2}$, $L_4: \dfrac{x}{2}=\dfrac{y-3}{1}=\dfrac{z-3}{2}$ then the minimum area of quadrilateral ABCD is ___

2. Let (x, y, z) is three-dimensional point which satisfies $xyz^4 = 16$ where (x, y, z > 0) its minimum distance from origin is λ, then λ^2 is ___

3. If $[\vec{a}\ \vec{b}\ \vec{c}]=2$ and $\vec{a}\cdot\vec{b}=2, \vec{c}\cdot\vec{a}=5$, then $\vec{a}$ is equal to $x\vec{b}+y\vec{c}+z(\vec{b}\times\vec{c})$ (where $\vec{b}$ & $\vec{c}$ are two non-coplanar orthogonal unit vectors) then (x + y − z) = ?

SUBJECTIVE PROBLEMS

1. If $\vec{a}=\hat{i}+\hat{j}+\hat{k}$ and $\vec{b}=\hat{j}-\hat{k}$, find a vector $\vec{c}$ such that $\vec{a}\times\vec{c}=\vec{b}$ & $\vec{a}\cdot\vec{c}=3$.

2. Find the points on line $\dfrac{x+2}{3}=\dfrac{y+1}{2}=\dfrac{z-3}{2}$ at a distance of 5 units from the point P(1, 3, 3).

3. Find the distance of the point (−2, 3, −4) from the line $\dfrac{x+2}{3}=\dfrac{2y+3}{4}=\dfrac{3z+4}{5}$ measured parallel to the plane 4x + 12y − 3z + 1 = 0.

THANKS!

Keep smiling!

Visit Us: https://www.mathsarc.com

ANSWER KEY – Vector & 3D

SINGLE OPTION CORRECT

1. C   2. C   3. D   4. A
5. B   6. B   7. D   8. A
9. B   10. C   11. B   12. D
13. C   14. A   15. D   16. C
17. D   18. B   19. A   20. C
21. D   22. C   23. D   24. C

MULTI OPTIONS CORRECT

1. A, C   2. A, C, D 3. A, D   4. A, B
5. A, D   6. B, C

INTEGER TYPE

1. 3   2.   3. 5

SUBJECTIVE

1. $\vec{c} = \dfrac{1}{3}\left(5\hat{i} + 2\hat{j} + 2\hat{k}\right)$ 2. (–2, –1, 3) or (4, 3, 7)   3. 17/2

ABOUT THE AUTHOR

Ramesh Chandra, a proud alumnus of Jawahar Navodaya Vidyalaya (JNV) and IIT Kanpur has always had a consistent academic record. His love towards Mathematics and passion towards teaching made him join FIITJEE as a teacher where he worked for almost 3years and 10 months during which he presented several papers for their A.I.T.S. in various groups, i.e. IIT JEE - mains, IIT JEE - Advanced, BITSAT, KVPY, and NTSE. Mr. Chandra currently works in a reputed education industry in Pune, Maharashtra and having teaching experience of 10+ years in education industries. His previous published books are 'PERMUTATION AND COMBINATIONS', "MATHEMATICS THE FIRST STEP" & Available at Mathsarc Education, Flipkart, Amazon and Notion Press etc.

Mr. Ramesh Chandra (AIR – 111)

B. Tech IIT Kanpur (Mechanical Engineering)

10 Years Mathematics Teaching Profession

- 4 year at FIITJEE East Delhi Centre — Associate Professor A - 1
- 1 year Mathsarc Education — Online Classes - Mathematics
- 1 Year at Atharva classes pune — Senior Professor
- 4 Year at Bakliwal Tutorials pune — Senior Professor

Now Director at VALUED CLASSES PUNE

BOOKS PUBLISHED

Buy at Amazon / Flipkart

Permutation and Combinations

(1) Amazon.in – Click Here
(2) Amazon.com – Click Here
(3) Flipkart.com – Click Here

E-Book at
   Kobo – Click Here

Buy at Amazon / Flipkart

Mathematics the First Step

(4) Amazon.in – Click Here

(5) Amazon.com – Click Here

(6) Flipkart.com – Click Here

E-Book at
Google Books – Click Here

Kobo – Click Here

Kindle – Click Here

iBooks – Click Here

THANKS!

Keep smiling!

Visit Us: https://www.mathsarc.com

Email Id – info@mathsarc.com